Popular Instructions on the Calculation of Probabilities

LAMBERT ADOLPHE JACQUES QUETELET
EDITED AND TRANSLATED BY
RICHARD BEAMISH

CAMBRIDGE
UNIVERSITY PRESS

CAMBRIDGE
UNIVERSITY PRESS

University Printing House, Cambridge, CB2 8BS, United Kingdom

Published in the United States of America by Cambridge University Press, New York

Cambridge University Press is part of the University of Cambridge.
It furthers the University's mission by disseminating knowledge in the pursuit of
education, learning and research at the highest international levels of excellence.

www.cambridge.org
Information on this title: www.cambridge.org/9781108064439

© in this compilation Cambridge University Press 2013

This edition first published 1839
This digitally printed version 2013

ISBN 978-1-108-06443-9 Paperback

POPULAR INSTRUCTIONS

ON

THE CALCULATION

OF

PROBABILITIES.

TRANSLATED FROM THE FRENCH OF

M. A. QUETELET,

PERPETUAL SECRETARY OF THE ROYAL ACADEMY OF BRUSSELS, CORRESPONDING
MEMBER OF THE INSTITUTE OF FRANCE, OF THE ROYAL ASTRONOMICAL SOCIETY
OF LONDON, OF THE ROYAL ACADEMIES OF BERLIN, TURIN, ETC.

TO WHICH ARE APPENDED

NOTES,

BY

RICHARD BEAMISH, ESQ., C.E., F.R.S., &c.

Mundum numeri regunt.

LONDON:

JOHN WEALE, ARCHITECTURAL LIBRARY,
HIGH HOLBORN.

1839.

EDITOR'S PREFACE.

I HAVE been induced to undertake the translation of this little Work, during my few leisure hours, from the aid which it has afforded me in exhibiting the valuable application of numbers to my pupils, and I am now tempted to publish it, with the omission of the account of a lottery proposed in the Netherlands, in the hope that it may prove equally serviceable to others.

The doctrine of chances has already secured important benefits to the public, in the establishment of Assurance Societies for life, health, and property, and is every day becoming more valued, in its application to the natural and moral sciences. By destroying that arbitrary power with which opinion had been invested, it leads directly to an humble observation of nature; to an acknowledgment of the fixedness of her laws; and to a conviction of the evils which arise from a disregard of her ordinances. By correcting the indefinite phraseology which is found to pervade all inquiries not embraced by those sciences, to which the direct application of numbers has been considered necessary, it tends to give regularity and symmetry to every pursuit, and to demonstrate, that the mean results obtained from the concurrence of the same circumstances, are as constant, as the numerical values of the natural philosopher, when determined by the repetition of the same experiments.

To the young men emerging from our public schools or colleges, the "Doctrine of Probabilities" offers many important practical advantages; impressed, as they often are, with an exalted idea of their knowledge of the world and its temptations, and undeceived only when they have become the victims of those who have made this subject their especial study, as applied to games of chance.

Mr. De Morgan, in his admirable Work on Probabilities, most justly observes, "that even those who derive absolute gratification from the

excitement which gaming affords, would hesitate
in their indulgence, were they convinced that
the contest in which they had engaged was so
unequal, that the ruin of their fortune, and too
often of their character, depended only on the
time devoted to the struggle. Absolutely speaking, young persons are not thoughtless with
respect to dangers of which they know the
risks. The ill success of others does not deter
them, because they attribute it to fortune; and
because they have superstitions hanging about
them with respect to luck, which are tolerably
prevalent in all ages. They think that they
are *trying their luck*, as the phrase is; but if
they could be convinced, that it is *not* their
luck which they are trying, but only a *fraction
of it*, their opponent having the rest in his
pocket, they would shew themselves in this, as
in other matters, averse to risks in which it is
more than an even chance against them."

Every day's experience shews how much society is yet tinged with the superstitions of a
barbarous age: the mass resting still satisfied
with the indefinite or deceptive results of a
natural perception, rather than seeking to obtain a rational conviction through the medium
of numerical calculations.

" Let us call to mind," observes M. De la
Place, in his Essai Philosophique sur les Pro-
babilités, " that formerly, and at a period not
far distant, an extremely wet or dry season,—
a comet, drawing after it a long tail,—eclipses,—
aurora boreales, and generally, all extraordinary
phenomena, were regarded as nothing but signs
of the wrath of Heaven. To turn from them
its baneful influence, Heaven was invoked. No
prayer was offered, however, to suspend the
course of the sun and planets, observation
having early declared the inutility of such
prayers; but as the former phenomena appeared and disappeared after long intervals, they
seemed to be contrary to the order of nature;
it was supposed that Heaven, irritated by the
crimes of earth, sent forth those prodigies to
announce its vengeance. Thus, the long tail
of the comet, in 1456, spread terror throughout
Europe, already alarmed by the rapid success
of the Turks in the eastern empire *. This
phenomena, after four of its revolutions, excites
amongst us a very different interest. The knowledge of the laws of the system of the world,

* " Pope Calixtus ordered public prayers to be said all
over Christendom, in which he exorcised the comet and
the Turks."—*Encyclopedia Britannica.*

acquired in the interval, has dissipated these
infantine fears, derived from an ignorance of
the true relation of man to the universe; and
Halley, having recognized the identity of this
comet, with that of the year 1531, 1607, and
1682, announced its next return for the end
of 1758 or the commencement of 1759." In
1757, Lalande, Clairaut, and Madame Lapaute,
wife of a celebrated watch-maker and astro-
nomer, undertook to calculate the disturbing
effects of certain of the planets on the movement of this comet: and the result was, that
in November 1758, the expected arrival of the
comet was predicted for the 13th April, 1759;
it was added, however, that from the imperfec-
tion of the method of calculation, there might
be an error of a month. It was actually observed December 25th, 1758, and it reached
its perihelion on the 13th March, 1759. The
re-appearance of this comet in 1835, as subsequently predicted, has further tended to con-
firm one of the grandest discoveries ever made
in science.

" Finally," says La Place, " if it be observed
that in those things which cannot be submitted
to direct calculation, the doctrine of Probabilities offers the most certain approximation to

guide us in our judgments, and the best guarantee against those illusions that so often deceive us; it will be admitted, that there is no
science more worthy of our meditations, and
none that would render more service, were it
introduced into a system of public instruction."

In an Appendix, I have given Tables which
will be found not devoid of interest: some derived from the more recent work of M. Quetelet, " Sur l'Homme et le Développement de
ses Facultés;" and others from Parliamentary
Papers. That " on the Sickness, Mortality,
and Invaliding among the Troops in the West
Indies," has afforded valuable extracts ; as has
also the Second Report of the Poor Law Commissioners, in connexion with M. Quetelet's
Observations and Tables relative to illegitimate
births.

Those relating to crime, in England and Wales,
and Ireland, will be found compared with some
given by M. Quetelet.

To Sir W. Herschel, Mr: De Morgan, Mr.
Edmonds, and other labourers for moral and
intellectual advancement, I shall be found also
indebted.

DODDERSHALL PARK, AYLESBURY,
     JANUARY, 1839.

AUTHOR'S PREFACE.

This little Work, which I consign to the public, is the summary of lectures, given for many years by me at the Museum of Brussels, as an introduction to my course of Physics and Astronomy. It seems to me that the calculation of probabilities, unfortunately too much neglected, ought according to the present state of knowledge, to serve as a foundation for the study of all the sciences, and particularly for that of observation. The greater portion of our knowledge rests in effect only on probabilites more or less strong, which are appreciated vaguely by the vulgar, and as it were by instinct; but which the philosopher, or at least the man who aspires to merit that title, ought to know how to appreciate according to certain rules.

Almost all our prejudices arise from the habit of pronouncing on simple appearances, but which will not bear a rigorous examination. Wherever things can be expressed by numbers, numbers have been taken as the guide; ceasing to dispute by counting, we have preferred facts rather than words.

The calculation of probability, which has hitherto had for its object the consideration of games of chance only, takes now a more elevated flight; bringing its light to the statesman, to regulate elections, and to examine the modes of arranging tribunals in the most advantageous manner; guiding the steps of the inquirer in his researches on births and deaths; fixing the basis of assurance societies; casting a new light on the system of our universe; and giving birth to statistics, that formidable arsenal from whence the orator, in mounting the rostrum, may provide himself with arms the most perfect.

The title of this corpuscule sufficiently shews that I have not written it for the learned, who may refer with more success to the works of La Croix, Parisot, the illustrious La Place, &c., to whom I am myself much indebted.

The Lessons Nos. XII. and XIII. are for the greater part extracts from the excellent introduction to the Statistical Researches on the City of Paris, which is due to the most distinguished geometrician of the age : the rule of the lesser square, which had scarcely been employed, save by astronomers versed in the knowledge of the higher mathematics, is there applied with a clearness which renders it available to observers the least accustomed to calculations.

I ought to state, that I suppose a knowledge of the first rules of arithmetic to be obtained, a knowledge which, according to the actual state of instruction in Belgium, may be reasonably expected from all who can read and write. I shall esteem myself most happy if this little Essay renders any service, by drawing attention to a branch of mathematics eminently in harmony with the progress of science.

SIGNS EMPLOYED IN THIS WORK.

\+ *Plus.* Sign of addition.

— *Minus.* Sign of subtraction.

× *Multiplication.* This is sometimes replaced by a point; thus 3×4 is differently written $3\cdot4$, to indicate that 3 is to be multiplied by 4: the numbers 3 and 4 are the two *factors* of the product 12. When the *same* factor is repeated many times in the same product, a more simple notation is employed, thus 3^2 represents 3×3, and 5^3 represents $5 \times 5 \times 5$.

To indicate that one quantity should be divided by another, the first is written above the second, the numbers being separated by an horizontal line, thus $\frac{15}{6}$ shews that 15 is to be divided by 6.

b

Numbers written in the above manner, are placed under the form of a *fraction;* 15 is the numerator, and 6 is the denominator.

The square of a number is the product of that number multiplied by itself; for example, 25 is the square of 5; 36 is the square of 6. The square root of a number is such a quantity that in multiplying it by itself it will reproduce the proposed number: for example, 5 is the square root of 25; 6 is the square root of 36. The symbol of the square root is $\sqrt{}$: thus $\sqrt{25}$ means, that the square root of 25 is to be extracted; in like manner $\sqrt{36}$ indicates 6, the square root of 36.

To shew that two quantities are equal, they are separated by the sign $=$.

CONTENTS.

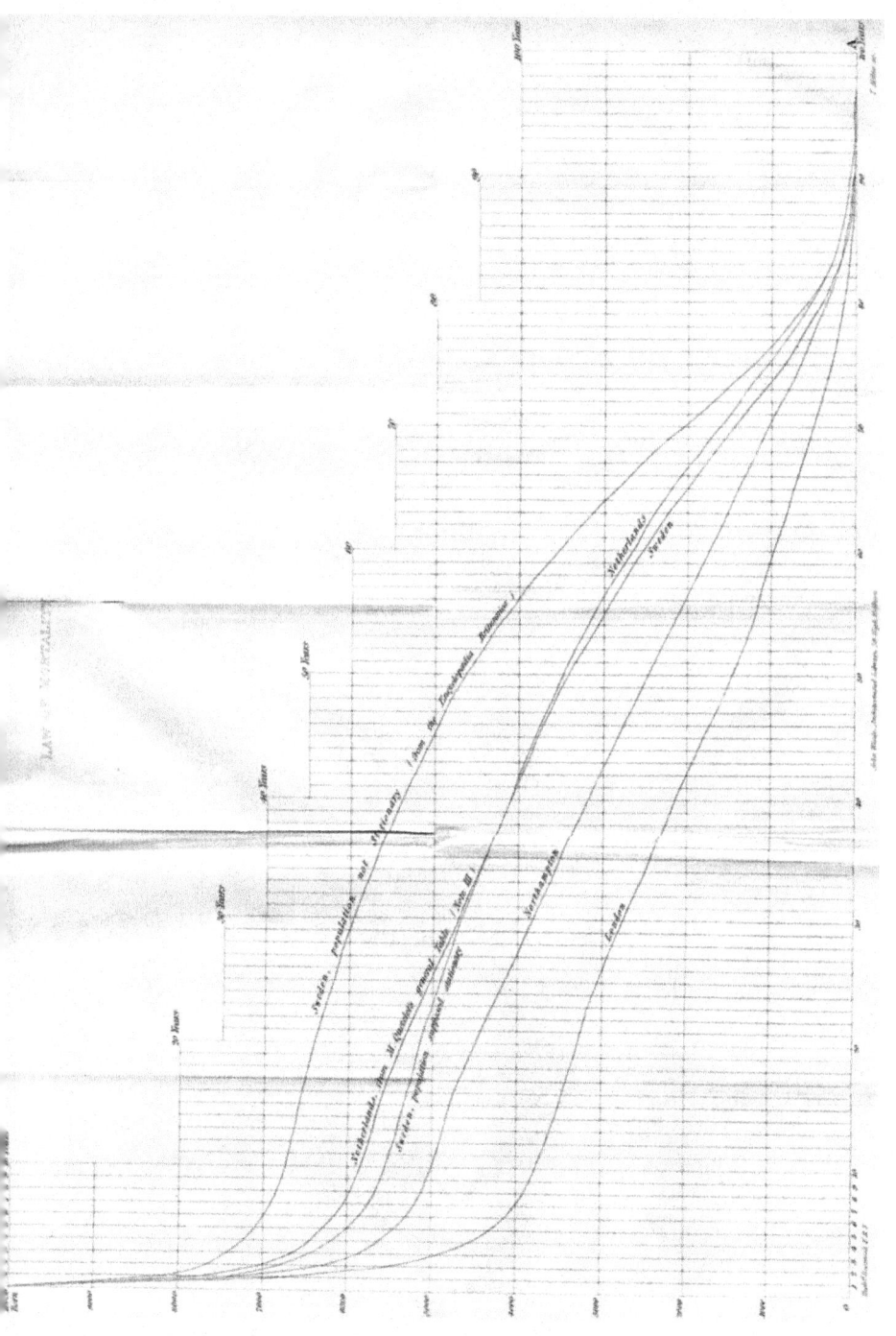

The material originally positioned here is too large for reproduction in this reissue. A PDF can be downloaded from the web address given on page iv of this book, by clicking on 'Resources Available'.

POPULAR INSTRUCTION,

ETC.

LESSON I.

ON CERTAINTY AND PROBABILITY.

WHEN different circumstances give rise to an event, they are called the *changes* of the event.

Example.—The drawing of one number in a lottery offers ninety chances; hence ninety different numbers may lead to the expected event.

In the throwing of some one point, an ordinary die presents six chances, as it must fall on one of its faces.

When the nature of the event which we hope for be declared, two sorts of chances arise, some favourable, others contrary or unfavourable to the event.

Example.—In a game with thirty-two cards, as in piquet, the chances of drawing a figured or court card are twelve, that is to say, as many as there are court cards, and twenty chances against.

Observation.—When all the chances are favourable to an expected event, their combination constitutes certainty.

Example.—Suppose an urn to contain three white balls, and you desire to draw one white ball; all the chances being favourable, we call the attainment of the proposed object a certainty.

Observation.—When there are only a certain number of favourable chances to the accomplishment of an event, the event is said to be *probable.*

Example.—If an urn contain three balls, one white and two black, the drawing of the white ball is probable; in three chances there is only one favourable.

The chances of drawing a king in a game with thirty-two cards, is again *probable;* out of thirty-two chances there are but four favourable, one or the other of the four kings may be drawn.

All events are not equally probable, and their different degrees of probability will be measured by the greater or less number of favourable chances.

Example.—If an urn contain twenty white balls and five black; the drawing of a white ball offers more favourable chances than that of a black, and we say that it is more probable than the other.

The same in a game of thirty-two cards, the drawing of a figured or court card is less probable than that of any other.

The calculation which teaches you to discover the degrees of probability of an event, is called the conjectural calculation, or calculation of probabilities.

In *some* cases the number of chances of an event are *limited,* and calculable, as in most of the games of Hazard; the probability of an event may generally be estimated with considerable facility, as we shall see in what follows.

In *other* cases the number of chances are *unlimited;* as in most of the sciences of observation.

We ought then to *estimate* the probability of the event, by means of a certain number of the chances which we obtain by experiment.

The consideration of this species of probability will form the second part of this essay: to give an example however of its existence, suppose an urn to contain an indefinite number of balls, the colours of which are unknown; and you find, that in a great number of trials, white balls only are drawn, and you are asked what the probabilities are that the urn contains only balls of this one colour.

You answer, *that it is probable* the urn contains only white balls; this assertion cannot however be considered as a certainty; for it may be, that one or several black balls, not yet drawn, remain in the urn. Thus, although the sun has been seen to rise with regularity for millions of years, yet it can only be *considered as probable* that this planet will again rise to-morrow: for some law may exist in nature not yet manifested, and which may prevent the sun from rising to-morrow.

We have not perhaps been in a condition to examine all the possible chances. We conceive, nevertheless, that there are probabilities so strong as to admit of their being considered as nearly certainties.

Example.—To the case where the probability of seeing the sun rise to-morrow, or of drawing a white ball from an urn, which, after a considerable number of trials has only yielded white balls, may be added also the probability of a man in health and strength living five minutes. There is little difficulty in receiving the event as certain, notwithstanding the fact that men, who had given promise of living many days, had been cut off by sudden death.

We always regard events as dependent on the causes which produce them ; and *chance* can only be considered as the effect of our ignorance of those causes.

We say, that a grain of dust, that a simple molicule of air or vapour floats at *hazard*, notwithstanding that the curve described is, as observed by the illustrious De la Place, regulated with as great certainty as the orbits of the planets; there is no difference between them but what is found in our ignorance.

QUESTIONS TO LESSON I.

What do you term the chances of an event?

What do you term the favourable and unfavourable chances of an event?

What is certainty?

What is probability?

Give examples of certainty and of probability; are there different degrees of probability?

What is the conjectural calculation of probabilities?

Are the number of chances of an event always limited and calculable?

When the number of chances of an event is unlimited, can you estimate still the probability of that event?

Are probabilities ever so great as to permit of their being considered as certainties?

How ought we to view that which we call *hazard?*

LESSON II.

ON MATHEMATICAL PROBABILITY.

In cases where all the chances of an event are equally possible, " the Mathematical Probability is estimated by dividing the number of favourable chances by the total number of chances."

If an urn contain 3 white balls and 2 black, there are three favourable chances out of five for drawing a white ball: and we say that the mathematical probability of the expected event is $\frac{2}{5}$. In like manner, the mathematical probability of drawing one figure, or court-card, in a game with 32 cards, is $\frac{12}{32}$; since among 32 cards there are 12 figured ones.

The probability, *unfavourable* to an expected event, is estimated in a similar manner; viz. by dividing the number of unfavourable chances by the total number of chances.

In the preceding examples the probabilities offered to the two expected events are $\frac{2}{5}$ and $\frac{20}{32}$.

Generally, each unexpected event gives rise to two opposite probabilities, viz. that the event shall occur, and that it shall not; and the sum of these two probabilities is equal to unity.

The probability of getting a figure card in a game with 32 cards, is $\frac{12}{32}$; the probability against it is $\frac{20}{32}$, and the sum of these probabilities is $\frac{12}{32}$ plus $\frac{20}{32}$ or 1.

It is then only necessary to subtract from 1 the mathematical probability favourable to an event, to obtain the probability against it.

The mathematical probability of an event, ought, after what has been said, to be expressed by a proper fraction, since the number of favourable chances cannot surpass the total number of chances.

It is obvious, that the greater the number of favourable chances in proportion to the total number of chances possible, the stronger will be the probability of the event occurring.

Example.—The probability of $\frac{12}{32}$ is greater than that of $\frac{4}{32}$; the former being the mathematical probability of obtaining a figure card at a game of 32 cards; and the latter that of one of the four aces.

When all the chances become favourable the result is *certainty*, and the numerator of the fraction becomes equal to the denominator, so that *unity is the symbol of certainty.*

When it is required to compare two mathematical probabilities, they must be reduced to a common denominator.

Example.—If it were required to know which were the more probable, the throwing of an ace with a die having six faces, or the drawing of a figure card in hearts, in a game with 32 cards. There will be for the first probability, $\frac{1}{6}$, and for the second, $\frac{3}{32}$; these fractions being reduced to a common denominator give $\frac{32}{192}$ and $\frac{18}{192}$. The first event is therefore more probable than the second.

The great defect (from habit) of estimating the probabilities of uncertain events, causes gross mistakes in estimating their value; it is always requisite to obtain a term of comparison, which may serve to rectify our judgments: the most

simple means seems to be, to conceive the favourable and unfavourable chances to be represented
numerically by white and black balls, which may
be contained in an urn: the accomplishment of
an expected event may be assimilated in this
manner to the drawing of a white ball.

Example.—What is the probability of throwing
an ace with a die having six faces? As there is
but one chance out of six, the probability will be
$\frac{1}{6}$, the same as that of drawing a white ball from
an urn which contains 6 balls, viz. one white
and 5 black; or again, What will be the probability of getting a king in a game with 32 cards?
As there are four chances favourable out of 32,
the probability will be $\frac{4}{32}$, the same as that of
drawing one white ball from an urn which contains 32 balls, viz. 4 white and 28 black.

Observe, that the increase and decrease of the
number of favourable chances, and the total number of possible chances, in the same proportion,
does not alter the probability, it remains the same;
thus, in place of the probability $\frac{4}{32}$, we may say
the probability $\frac{1}{8}$, which will be found an equi-
valent.

The probability of drawing one white ball from
an urn which contains 32 balls: thus, 4 white
and 28 black, is precisely the same as drawing
one white ball from an urn which contains 8 balls,

viz. one white and 7 black: this means of sim-
plification, depending on the properties of fractions, will be found often useful.

The mode suggested of estimating the value of
probabilities offers, however, an inconvenience,
inasmuch as it would be difficult to appreciate the
amount of value which may be attached to a probability, to permit of its being classed among
those which we have been habituated to consider
as certainties.

The best term of comparison seems to be the
probability of living still through a certain space
of time. This measure will be more appreciated
in consequence of the value which we attach
generally to life, than any other more precise
mode that we are in the habit of using.

If we cast our eyes on a table of mortality,
on that of the southern provinces of the Netherlands for instance, we shall see, that out of 51,956
persons who had attained to the age of 20, the
tenth part ceased to exist at the termination of
7 years.

Thus: at this age the probability of dying in
the space of 7 years is $\frac{1}{10}$, that is to say, something less than the probability of getting a king
at the first cut with 32 cards. By making similar
calculations for the periods at which the $\frac{1}{100}$ the
$\frac{1}{1000}$ of these 51,956 young persons cease to exist,
the following Table is formed, to which we shall
have occasion to refer.

TABLE.

Probability	of dying before
$\frac{1}{10}$	7 years.
$\frac{1}{100}$	8 months.
$\frac{1}{1,000}$	25 days.
$\frac{1}{10,000}$	60 hours.
$\frac{1}{100,000}$	6 hours.
$\frac{1}{1,000,000}$	36 minutes.
$\frac{1}{10,000,000}$	4 minutes.
$\frac{1}{100,000,000}$	22 seconds.
$\frac{1}{200,000,000}$	1 second.

It must be observed that these results can be only received as of general application, not to individuals who may be in health, but rather as the probability that the new born infant, if he attain the age of 20 years, will die before a certain time designated in the Table.

Example.—Supposing it were required to know what the probability would be of the letters composing the word " Constantinople," after being thrown in the air should arrange themselves so as to recompose the same word. We know by calculations, which cannot be given here, that these 14 letters may be arranged more than 87,000,000,000 different ways, and yet only produce the same word 24 times: we have then for the probability of the throw $\frac{24}{87,000,000,000}$, in other words, a probability less than that of dying in the space of a second at the age of 20 years. It may therefore be regarded as certain, that the new born infant, should he attain the age of 20 years, may calculate on a second of existence ; it may be also deemed certain that the word " Constantinople" will not be produced by the first throw of the letters which compose it. We may further admit as certain, that the newly born infant, should he attain the age of 20 years, will live still 36 minutes ; this event having for the adverse chances $\frac{1}{1,000,000}$. Thus we shall regard as nearly certain those events which have their adverse chances less than one millionth : such would be the drawing a white ball from an urn containing one million of balls ; viz. 1 black and 999,999 white balls.

QUESTIONS TO LESSON II.

What is the mathematical probability of an event?

Give examples in the calculation of mathematical probabilities.

How do you calculate the probability contrary to an event?

What is the value of the two probabilities of an event?

Can the mathematical probability of an event be greater than unity?

How do you represent mathematically the various degrees of probabilities?

When is there certainty?

What is the symbol of certainty?

How do you place probabilities in a condition to be compared?

Is it not desirable to have a comparable term, in order to obtain a just idea of the relative magnitude of probabilities?

Can you simplify the expression of a mathematical probability?

What means would you employ to obtain a measure of the absolute value of probabilities?

In what case do you habitually consider an event as certain, which is only probable?

LESSON III.

ON SIMPLE AND COMPOUND CHANCES OR PROBABILITIES.

WHEN an event depends on many events independent of one another, it is said to be compound.

Example.—The throwing successively of the *ace* and *deuce* with ordinary dice having six faces, is a compound event, which depends on two events altogether independent the one of the other, viz. the throwing of the *ace* and that of the *deuce*.

The events independent of one another are *simple* events.

In the above example the attainment of the compound event depended on two simple events, which are the throwing of the *ace* and that of the *deuce*. We have before seen (Lesson II.) that the probability of a simple event, called simple probability, is estimated by dividing the number of favourable chances by the total number of chances.

A compound probability or the mathematical probability of a compound event, is obtained by the product of the probabilities of all the simple events on which the compound event may depend.

Example.—The probability of throwing the *ace* and then the *deuce*, with ordinary dice, will be calculated as follows:—The probability of throwing an *ace*, with one die, is $\frac{1}{6}$; so is that of a *deuce* $\frac{1}{6}$: The product of these two particular probabilities, or $\frac{1}{6} \times \frac{1}{6} = \frac{1}{36}$ forms the probability of the looked for event: this result will, by a little attention to the following Table, No. 2, be soon recognized.

TABLE, No. 2.

1 and 1	2 and 1	3 and 1	4 and 1	5 and 1	6 and 1
1 .. 2	2 .. 2	3 .. 2	4 .. 2	5 .. 2	6 .. 2
1 .. 3	2 .. 3	3 .. 3	4 .. 3	5 .. 3	6 .. 3
1 .. 4	2 .. 4	3 .. 4	4 .. 4	5 .. 4	6 .. 4
1 .. 5	2 .. 5	3 .. 5	4 .. 5	5 .. 5	6 .. 5
1 .. 6	2 .. 6	3 .. 6	4 .. 6	5 .. 6	6 .. 6

What is the probability of drawing successively the ace, king, queen, and knave of hearts, in a game with 32 cards, taking care each time to replace, in the pack, the card drawn: The compound event depends here on 4 simple events, which has each its probability equal to $\frac{1}{32}$; the compound probability becomes then $\frac{1}{32} \times \frac{1}{32} \times \frac{1}{32} \times \frac{1}{32} = \frac{1}{1,048,476}$. Thus, the expected event is less probable than the supposition that the newly born infant will die precisely in 36 minutes which follow the instant he has attained his 20th year.

We have already seen, that events having so little probability, are considered as *extraordinary*, or to say the least, uncommon.

C

Example.—What is the probability of throwing head, 20 times following, at a game of pitch and toss? We have here but 2 chances, of which 1 is favourable ; the simple probability is, therefore, $\frac{1}{2}$; the compound probability will be $(\frac{1}{2})^{20}$, that is to say, less than $\frac{1}{1,000,000}$; thus, the throwing head 20 times following, is another example of extraordinary events.

Example.—A fact which, without being extraordinary, has not in itself much probability, is transmitted to us by twenty witnesses, so that the first transmitted to the second, the second to the third, and so on, it is required to determine the probability of the fact, supposing that the correctness of each witness is equal to $\frac{9}{10}$ of the whole truth, (allowing 1 fault out of 10 evidences). We shall have for the required probability $(\frac{9}{10})^{20}$, that is to say, less than $\frac{1}{8}$. The truth will be less probable than the drawing of an ace in a game with 32 cards.

" We cannot better compare this diminution of probability," observes M. De la Place, " than the obscuration of the clearness of objects, by the interposition of numerous pieces of glass ; a number of small pieces being sufficient to obstruct the view of an object, through which a single piece would permit you to see distinctly.

" Historians do not appear to have attended sufficiently to this degradation of the probabilities of

facts, when viewed through a number of successive generations; many historical events reputed
certain would appear, at least, *doubtful*, if submitted to this test."

QUESTIONS TO LESSON III.

What is a compound event?
What is a simple event?
How do you designate a simple and compound probability?
How do you calculate a simple and compound probability?
Give an example of the calculation of a compound proba-
bility in the throwing of dice, and in the drawing of cards.
Give an example of the calculation of a compound proba-
bility in estimating evidence.
How do you represent the diminution of probability?

LESSON IV.

ON RELATIVE PROBABILITIES.

ALTHOUGH we have only, as yet, considered two
sorts of cases, many more may suggest themselves.

Example 1st.—If it were required to determine
the probability in a game with 32 cards, of getting
a figure card, or an ace; there would be three
possible events, of which the probabilities are:

$\frac{12}{32}$ for a figure card.

$\frac{4}{32}$ for an ace.

$\frac{16}{32}$ for any other card, not being either ace or
figure.

Example 2*nd.*—If in an urn there be 20 balls, viz. 8 white, 4 black, 3 red, and 5 green : there would arise, in the drawing, four sorts of chances, of which the several probabilities will be :

$\frac{8}{20}$ for a white ball.

$\frac{4}{20}$ for a black ball.

$\frac{3}{20}$ for a red ball.

$\frac{5}{20}$ for a green ball.

The sum of these probabilities, as in the preceding example, ought to be equal to unity.

Let us take a third example.

Example 3*rd.*—What are the probabilities of throwing with two dice, 7, 8, 9, or 10 points?

7 points may be obtained by 6 different combinations with two dice, as may be seen by the Table (Lesson 3d, Table No. 2). Thus 1 and 6, 2 and 5, 3 and 4, 4 and 3, 5 and 2, and 6 and 1, may be thrown.

In a similar manner it will be found, that 8 points may be thrown in 5 ways, 9 points in 4 ways, and 10 points in three ways only. Hence, for the probabilities of this description of chances, observing that there are 36 throws equally possible, we may obtain

$\frac{6}{36}$ for   7 points

$\frac{5}{36}$ —   8 —

$\frac{4}{36}$ —   9 —

$\frac{3}{36}$ —   10 —

$\frac{18}{36}$ for all other chances.

It is only usual to consider the *absolute* probability of events; sometimes, however, it is desirable to know the *relative* probability which one thing may bear to others.

Example.—What are the relative probabilities of drawing a figure card or an ace in a game with 32 cards ?—We have already seen, that for the one there are 12 chances, and for the other, 4 : disregarding then all the other chances, we consider only the 16 chances, one of the probabilities will be $\frac{12}{16}$, and the other $\frac{4}{16}$; the first being therefore triple of the second.

In the second example, the relative probabilities of drawing a white or a black ball will be $\frac{8}{12}$ and $\frac{4}{12}$, the drawing of the balls of other colours not entering into the computation.

In the third example, the probability of obtaining 7 points with two dice is double that of obtaining 10 points. There being in effect 6 chances in favour of the first event, and only 3 in that of the second out of 9 chances in all, omitting the other chances, which cause neither gain nor loss. An event often depends on many chances which are not equally possible, it is then necessary to determine the respective probability of these chances, and the probability of the expected event will be compounded of their sum.

Example.—Two persons play together ; one bets that he will, at the first trial, with two dice,

throw 7, 8, 9, or 10 points : as we have already
seen, the probabilities respectively of these events
are $\frac{6}{36}$, $\frac{5}{36}$, $\frac{4}{36}$, and $\frac{3}{36}$; the probability that he will
win, is $\frac{18}{36}$; the sum of these several probabilities,
and the probability against him will also be $\frac{18}{36}$.

Example.—The probability of drawing a figure
card or an ace in a game with 32 cards, should be
$\frac{12}{32}$ plus $\frac{4}{32}$ or $\frac{16}{32}$; that is to say, the sum of the
probabilities of the two descriptions of chances
which unite in favour of the event.

From what has been said, this valuable conclusion may be drawn, that it is necessary, when
any question is proposed, to examine whether the
looked for event be *simple* or *compound*, and
whether the chances to be calculated be all equally
possible, or not. Lastly, it will be necessary to
examine whether the probability be *absolute* or
relative; in other words, whether the favourable
chances are to be compared with all the chances,
or to some one in particular.

QUESTIONS TO LESSON IV.

Do you never consider but two sorts of chances?
Give examples which present a variety of chances.
How do you estimate a probability relatively to others?
Give examples of relative probabilities.
Are the chances of an event always equally possible?
How do you calculate the probability of an event of which
all the chances are not equally possible?

Give examples of the calculation of the probability of an event ·where all the chances are not equally possible.

What is the conclusion to be drawn from all that has been shown ?

LESSON V.

ON REPEATED EXPERIMENTS.

THOSE which are made successively under the same circumstances; as, for example, the drawing of cards in a game, and each time replacing them, or successive throws of the same dice.

Probabilities in relation to repeated experiments are calculated as compound events.

Example.—It is proposed with one die, having 6 faces, to throw the ace three times following; the mathematical probability will be $\frac{1}{6} \times \frac{1}{6} \times \frac{1}{6}$, or $\frac{1}{216}$: the expected event depends, in effect, on three simple events, which it will be necessary to multiply together.

Example.—It is proposed, in a game with 32 cards, to draw either a king or a queen three times following, taking care to replace each time the card' which had been drawn : the probability will be $\frac{1}{4} \times \frac{1}{4} \times \frac{1}{4}$: the event depending on three simple events, the probability of each being $\frac{1}{4}$, inasmuch as there will be 8 chances out of 32 in favour of each event.

Example.—What is the probability in two trials or tosses in the game of pitch and toss of obtaining first head, then tail ?—The probability will be $\frac{1}{4}$, that is to say, $\frac{1}{2} \times \frac{1}{2}$ the product of the probabilities of two simple events. The probability would have been $\frac{1}{2}$ if the conditions permitted a disregard for the order in which the throw came, whether tail or head ; in fact, there can occur but 4 compound events which have each a probability of $\frac{1}{4}$; viz:— throwing tail twice following, throwing head twice following, throwing tail and head, and throwing head and tail. Now, we have in our favour in this case, the probabilities of the last two events, *i. e.* $\frac{1}{4} + \frac{1}{4}$ or $\frac{1}{2}$. Observe that generally in the *repetition* of two experiments only four possible events present themselves, each experiment producing only two simple different events, which may be designated by the letters A and B. These events will be presented therefore in the following order :

TABLE, No. 3.

AA.   AB.   BA.   BB.

If we make a third trial, the number of possible events will be found to be doubled, and we should have 8 events. Thus, using the foregoing letters, we have

AAA.   ABA.   BAA.   BBA. for A.
AAB.   ABB.   BAB.   BBB. — B.

If we make a fourth trial, the number of possible

events will be again doubled, and we shall have
sixteen : thus :—

AAAA.   ABAA.   BAAA.   BBAA.⎫
AABA.   ABBA.   BABA.   BBBA.⎬ for A.
AAAB.   ABAB.   BAAB.   BBAB.⎫
AABB.   ABBB.   BABB.   BBBB.⎬ for B.

By following the same reasoning, it will be
found, that in making a fifth trial, the number of
possible events will be again doubled, and in a
similar manner ; hence the following table :—

TABLE, No. 4.

No. of Trials.	Possible Events.	
1 ...	2...	2
2 ...	4...	2×2
3 ...	8....	$2 \times 2 \times 2$
4 ...	16...	$2 \times 2 \times 2 \times 2$
5 ...	32...	$2 \times 2 \times 2 \times 2 \times 2$
6 ...	64...	$2 \times 2 \times 2 \times 2 \times 2 \times 2$
7 ...	128...	$2 \times 2 \times 2 \times 2 \times 2 \times 2 \times 2$
8	256....	$2 \times 2 \times 2 \times 2 \times 2 \times 2 \times 2 \times 2$
9 ...	512...	$2 \times 2 \times 2 \times 2 \times 2 \times 2 \times 2 \times 2 \times 2$
10 ...	1024...	$2 \times 2 \times 2 \times 2 \times 2 \times 2 \times 2 \times 2 \times 2 \times 2$

We see that the *number* of different compound
events become considerable when we attend to the
order in which the simple events present themselves, and which is found by multiplying successively 2 by itself (or that power of 2 designated by
the number of trials.—TR.)

Example.—Suppose then that you wish to know the probability of drawing in 4 trials, from an urn which contains 1 white and 2 black balls, at the 2 first trials, a white ball, and at the 2 last, a black ball. Always returning the ball drawn to the urn. On consulting the Table it will be found, that of 16 possible events, one alone is favourable, viz. AABB. Suppose A to designate the draught of the white ball, and B that of the black, then the probability of the event A is $\frac{1}{3}$, and that of B is $\frac{2}{3}$; from which the required probability will be $\frac{2}{3} \times \frac{2}{3} \times \frac{1}{3} \times \frac{1}{3} = \frac{4}{81}$. The contrary probability will be $\frac{77}{81}$, the two probabilities together producing unity.

Generally, if in repeated experiments, the attainment of an event A be looked for a certain number of times, and that of an event B also ; and further, if the order of occurrence be declared, the required probability will be a product which will involve the simple probability of the event A, as many times as that event shall be looked for, and so also of the probabilities which involve B.

Example.—What is the probability in a game with 32 cards of obtaining twice following a figure card, then three times following an ace ? The probability of obtaining 1 figure card being $\frac{12}{32}$ or $\frac{3}{8}$, and that of an ace being $\frac{4}{32}$ or $\frac{1}{8}$, we have for the required probability $\frac{3}{8} \times \frac{3}{8} \times \frac{1}{8} \times \frac{1}{8} \times \frac{1}{8} = \frac{9}{32,768}$.

Should the order be disregarded in which each event is to be presented, the number of the compound events will be found much reduced. Thus, for one experiment, we have only two possible events.

In the repetition of two trials, the event A may occur twice following, or only once, or it may not occur at all, thus:—

TABLE, No. 5.

AA.   AB.   BB.
BA.

In the repetition of three trials, the event A may occur three times, twice, once or not at all:

AAA.   AAB.   BBA.   BBB.
ABA.   BAB.
BAA.   ABB.

In the repetition of four trials, the event A may occur 4 times, 3 times, twice, once, or not at all:

AAAA.   AAAB.   AABB.   BBBA.   BBBB.
AABA.   ABAB.   BBAB.
ABAA.   BAAB.   BABB.
BAAA.   ABBA.   ABBB.
BBAA.
BABA.

It will be observed, that this is only another arrangement given to the former table. The analogy enables us to form the following table:

TABLE, No. 6.

Repeated trials.	Events.
1	2 .... 1+1

Repeated trials.		Events.		
2	....	3	....	2 + 1
3	....	4	....	3 + 1
4	....	5	....	4 + 1
5	....	6	....	5 + 1
6	....	7	....	6 + 1
7	....	8	....	7 + 1
8	....	9	....	8 + 1
9	....	10	....	9 + 1
10	....	11	....	10 + 1

Thus, where the order in which the simple events occur is indifferent, the number of possible events is equal to the number of repeated trials plus 1.

Example.—If from an urn containing 1 white and 2 black balls, it be required to know the probability of drawing in four trials, twice a white and twice a black ball, without determining the order in which they must be drawn, and always returning the ball drawn to the urn, it will be seen by reference to Table 5, that this event may occur in six ways; there will then be 6 times the probability that one only of these events shall occur, *i. e.* $6 \times \frac{4}{81}$ or $\frac{24}{81}$; the probability against is $\frac{57}{81}$.

Many different probabilities may be re-united; thus, in the preceding example, if it be proposed to draw in 4 trials at least twice a white ball, we have the probability which has been calculated plus the probabilities of drawing 4 or 3 white balls.

These sorts of calculations become exceedingly simple by the employment of Algebra.

Numerical calculations are much simplified when the probability of the accomplishment of an event is required, which shall occur once at least in a given number of repeated trials. As there is only against it the probability that the expected event does not occur at all; that chance is calculated and deducted from unity.

Example.—It is required to know the probability of throwing the ace at least once, with a die having six faces, and in three repeated trials; the probability of *not* throwing the ace is $\frac{5}{6}$, and of not throwing it three times following, is $\frac{5}{6} \times \frac{5}{6} \times \frac{5}{6}$, or $\frac{125}{216}$: as this is the only probability *against* the event, by deducting it from 1, there remains for the probability of the event proposed $\frac{91}{216}$, or a little less than $\frac{1}{2}$.

QUESTIONS TO LESSON V.

What do you designate repeated trials?

How do you calculate the probability in this description of trial?

Give examples of calculation of the mathematical probability in repeated trials.

How many possible events may arise by the repetition of two, three, or four trials, when the order of succession is attended to?

How many possible events may arise by the repetition of

D

any number of trials, when the order of succession is considered?

Illustrate what has preceded by an example.

How many possible events may arise by the repetition of two, three, four, or any number of trials, when the order of succession is not attended to?

Illustrate by an example.

How do you calculate the probability that a named event will occur at least once in a given number of trials?

LESSON VI.

OF SOME PARTICULAR CASES IN CALCULATING MATHEMATICAL PROBABILITIES.

IN considering the repetition of trials, we have supposed that the number of chances remain the same at each new trial; but this circumstance may not occur.

Example.—What is the probability that in two trials, a figure card be drawn in a game with 32 cards, supposing the card drawn not to be returned the first time? This may occur in two ways, by drawing a figure card at the first trial or at the second.

For the first, the probability will be $\frac{12}{32}$ or $\frac{3}{8}$. The second trial becomes useless, if the first be successful; it is not then certain that it ought to be made, and the probability is $\frac{5}{8}$; that is to say, it is equal to the probability that the first trial will not succeed. But if the second trial be required it will be with 31 cards, of which 12 are favourable to the attempt; the probability of drawing a figure card will be then $\frac{12}{31}$, if the drawing could be deemed certain, and it is $\frac{5}{8} \times \frac{12}{31}$, since it has but $\frac{5}{8}$ of probability. There is in this manner, for the expected event, a probability equal to $\frac{3}{8} + \frac{5}{8} \times \frac{12}{31}$.

An urn contains 2 white and 2 black balls, and 2 players, A and B agree (their eyes being bandaged) that the one who shall first draw a white ball, wins. They draw alternately, and A begins. It is required to know the chances each has in his favour, supposing that the balls drawn be not returned. At the first trial the probability of drawing a white ball is $\frac{2}{4}$ or $\frac{1}{2}$.

The second trial is not certain, its probability is $\frac{1}{2}$, that is to say, the probability that the player A will draw a black ball the first time: if the second trial be permitted, as there will remain in the urn 2 white balls and 1 black, the probability of drawing a white, will be $\frac{2}{3}$; and in this case it will be $\frac{1}{2} \times \frac{2}{3}$, or $\frac{1}{3}$, because the second trial is doubtful. The probability of drawing a black ball, however, will be $\frac{1}{2} \times \frac{1}{3}$ or $\frac{1}{6}$; which is also the probability of the third trial. But, as this third time the urn contains only white balls, there is a certainty of drawing one; it will be necessary then to multiply 1 by $\frac{1}{6}$ the probability of making the third trial.

Thus, the chances in favour of the player A are $\frac{1}{2}+\frac{1}{6}=\frac{2}{3}$, and in favour of the second player, $\frac{1}{3}$: these two probabilities united, equal 1, $(\frac{1}{2}+\frac{1}{6}=\frac{8}{12}$ and $\frac{8}{12}+\frac{1}{3}=\frac{36}{36}$ or 1.)

Let us take another example which may offer some difficulty. Supposing that two urns are placed before an individual, one containing 2 white balls and 5 black, the other containing 3 white and 1 black; what will be the probability of drawing a white ball from one of these urns? The probability that the person making the trial will draw a white ball from the first urn, depends on two events, viz. on the choice of the urn, and the drawing. The probability of the first urn being selected is $\frac{1}{2}$, and the probability of drawing a white ball is $\frac{2}{7}$; the probability of drawing a white ball therefore from the first urn will be $\frac{1}{2}\times\frac{2}{7}$ or $\frac{1}{7}$. The probability of drawing a white ball from the second urn will be also $\frac{1}{2}\times\frac{3}{4}$ or $\frac{3}{8}$. Thus the probability of drawing 1 white ball from one or other of the two urns, will be $\frac{1}{7}+\frac{3}{8}$.

We conclude by an example which will show how necessary it is to be guarded against first appearances in the calculation of probabilities. It may be generally supposed a matter of indifference whether you bet on drawing even or odd numbers from an urn containing a certain number of balls. However, by betting on the odd numbers there is always one chance more in your favour than on the even numbers; for example, if the urn contain but one ball, there is but one chance, and it will be favourable to him who bets on the odd numbers. If the urn holds two balls, a and b suppose, they may be drawn in the following manner:—

$$a, b, ab;$$

the two first being odd, the third even.

If the urn hold three balls, a, b, and c, they may be drawn in seven ways:

$$a, b, c, abc, ab, ac, bc;$$

the first four being odd, and the three last even.

If the urn contain 4 balls, a, b, c and d, they may be drawn in 15 different ways:

$$a, b, c, d, abc, abd, acd, bcd;$$
$$ab, ac, ad, bc, bd, cd, abcd;$$

the eight first being odd, and the seven last even.

Generally the probabilities will be as follow:—

TABLE, No. 7.

Number of balls.	Probability of drawing an odd number.	Even.
1	1	0
2	$\frac{2}{3}$	$\frac{1}{3}$
3	$\frac{4}{7}$	$\frac{3}{7}$
4	$\frac{8}{15}$	$\frac{7}{15}$
5	$\frac{16}{31}$	$\frac{15}{31}$
6	$\frac{32}{63}$	$\frac{31}{63}$
7	$\frac{64}{127}$	$\frac{63}{127}$

Number of balls.	Probability of drawing an odd number.	Even.
8	$\frac{128}{255}$	$\frac{127}{255}$
9	$\frac{256}{511}$	$\frac{255}{511}$
10	$\frac{512}{1023}$	$\frac{511}{1023}$

The mode of continuing the table is obvious; for in mathematical probabilities, to obtain an odd number the numerator of each fraction is equal to double the preceding numerator; and the denominator is double the denominator, less one.

QUESTIONS TO LESSON VI.

How do you calculate the probability in successive trials when the card drawn is not returned each time to the pack?

Give examples of similar probabilities.

How do you calculate the probability of drawing a white ball from two urns which enclose balls of different colours?

When at hazard, you draw from an urn which encloses a certain number of balls, is it more probable that you will draw odd rather than even numbers?

What are the probabilities for drawing even, and what for odd?

LESSON VII.

ON THE MANNER OF EXAMINING PROBABILITIES.

WHEN it is desirable to apply conjectural calculations, it becomes interesting to inquire how far the results of these calculations accord with those of experience. This species of inquiry has much occupied geometricians (and more particularly J. Bernouilli). We must be satisfied here to offer the conclusions at which they have arrived.

If one experiment only be made, the accordance with general experience cannot be known.

Calculation, in point of fact, offers but two probabilities, and experience determines the kind.

Example.—In throwing a die having six faces, and desiring an ace, there will be for the calculation of the probability of this event $\frac{1}{6}$, and against it $\frac{5}{6}$. When the experiment is made, however, the condition of the players is not found to conform to these calculations, inasmuch as the one will have gained and the other lost.

It is necessary to observe, that *in particular cases*, the calculation of probabilities will be found incorrect; but, when a great number of experiments are made, the accordance tends more and more to be re-established between the calculation and actual experience.

If an infinite number of trials could be made, the events will be only then distributed in the manner indicated by the calculation : in all other circumstances it is a mere chance whether this accordance shall exist, and the chances increase as the number of trials augment.

J. Bernouilli has found, that by multiplying sufficiently the number of trials, a probability may be obtained as nearly approximating to unity as can be desired, that the ratio of the repetitions of an event to the number of trials will involve its mathematical probability within the approximate limits that may be desired.

It is then exhibiting but little prudence to expose oneself to the chances of a hazard, where a great number of trials are not permitted.

Example.—Persons engaged in a lottery are under the circumstances above alluded to : the government, on the contrary, having many trials, ought to expect that the results will be found to accord very nearly with the calculations; also, as we shall have occasion to show, the profit which it obtains furnishes a value *which does not exceed* certain limits.

What has been said may be reduced to the three following propositions :

If the probability of an event exceed $\frac{1}{2}$, there is reason to suppose that the event will happen, rather than that it will not.

The more this probability augments, the more may expectation be increased.

Expectancy will be in proportion to this probability.

It will be found also, that in any number of experiments or trials, the most probable of the compound events will be that, where each simple event is repeated proportionally to its probability.

Example.—In drawing, by repeated trials, many cards in a game, the most probable recurrence of the events will be where the figure cards and the other cards not figured, are distributed in the proportion of 3 to 5, which is the simple proportion of each of these events, viz. $\frac{12}{32}$ and $\frac{20}{32}$.

When the various chances of a game are strictly of an equal possibility, as well by the construction of the instruments of distribution, as in the mode of using them, the passed events would have no influence on the future events.

It is the ignorance of this principle which leads so many to compromise their fortunes by embarking in lotteries.

If, after repeated trials, a marked frequency in the appearance of certain chances be observed, you will be justified in concluding that the constitution of the instrument, or the dexterity of him who employs it, is the cause.

Example.—If, in throwing a die, many times, having 6 faces, the *ace* presents itself oftener than the probability should indicate, it may be concluded that the die is loaded.

If, in pitch and toss, head comes down oftener than tail, it must be concluded, that in the construction of the piece, there exists a certain cause which favours the return of head.

Thus, observes M. de La Place, in the conduct of life, constant good fortune is a proof of skill, which should induce us to prefer employing those happy persons.

We have heard it stated, that during the first Spanish war, a corps of the French army which was employed in the siege of a town, dreaded the recurrence of Friday, as a day fatal to it, because the enemy killed or wounded then, more than on any other day of the week; a strong prejudice was in consequence established against Friday. Now after the siege it appeared that the artillery had been changed every day, and that those who did duty on Fridays were superior marksmen to the others. Superstitious ideas have often less foundation; from the impossibility of discovering the true cause under the circumstances in which we are placed, we are apt to attribute them to objects totally extraneous, not being willing to consider effects as existing without a cause.

QUESTIONS TO LESSON VII.

Will calculation, and experience agree, when only one trial is made?

Do they agree when an infinite number of trials are made?

Can you assign the number of trials, which would allow the results of calculation to approximate as much as may be desired to the results of experience?

Is it prudent to expose yourself to the chances of a hazard where a great many trials are not permitted?

What are the three principles to which you ought to have regard in the employment of probabilities?

What is the most probable of compound events in any number of repeated trials?

Have past events any influence on future events?

If, after repeated trials, a marked frequency is observed in the appearance of certain chances, what ought you to conclude?

Give some examples.

LESSON VIII.

ON MATHEMATICAL EXPECTATION.

LET two players be placed in such a position that neither shall have an advantage over the other; thus, when two persons make a bet, or play together, and that each has the same probability of gaining in his favour, it is obvious that justice demands that they hazard equal sums.

Example.—At the game of pitch and toss, the two players are in exactly similar positions; there can be in effect but two chances, and each has one in his favour; no motive for preference existing, these players ought to run the same risks and hazard the same sums.

If an urn contain 20 white and 20 black balls, and it be required to draw from it one ball of a specified colour, the probability of gaining for each player will be $\frac{1}{2}$; being, therefore, in the same condition, these players ought to run the same risks.

In the throwing of a die having six faces, six persons bet that each will throw a different face or number; these persons ought to hazard equal sums, for they are each in a position equally advantageous, having $\frac{1}{6}$ probability of gaining.

Should the probabilities of gaining not be the same, the players ought to hazard sums proportionate to these probabilities.

Examples.—If it be proposed to throw an ace with a die having 6 faces, the one who bets ought to hazard the fifth of that which the other hazards; he having, in fact, but 1 chance, while the other has 5. In order the better to explain this matter, suppose, as in the preceding example, each of the 6 players to hazard the same sum when betting on a different face; there would be then 6 sums equally hazarded; but it becomes indifferent to

the first player whether a single person be substituted for the other 5, provided he hazard, to
him alone, as much as he would have hazarded to
the 5 players.

Example.—In a game with 32 cards, a card is
drawn without its being returned, and a bet is
made that the card is a figure card; another person bets to the contrary. What sums ought to
be hazarded to render the bets equal? The pro-
bability of drawing a figure card is $\frac{12}{32}$ or $\frac{3}{8}$, and
that of not drawing one is $\frac{5}{8}$ or $\frac{20}{32}$; the sums
hazarded ought to be in the ratio of 3 to 5, similar
to the probabilities.

Let us suppose 8 equal chances, and 8 players,
each having one of these chances, or $\frac{1}{8}$ probability
of gaining, each ought to hazard the same sum,
1 shilling, for example; but *one* person may be
substituted for *three* of these players, paying their
expenses; and another person for the five, in paying equally their losses; the first ought then to
pay 3 shillings, and the second 5.

These sums are precisely in the ratio of the
chances, and the probabilities that the 2 players
have of gaining.

*Mathematical expectation, is the product of the
sum which one hopes to gain, by the probability
which one has of obtaining it.*

Example.—A bet is proposed, that at the first
cut a figure card, in a game with 32 cards, shall

E

be turned up, and 3 shillings are hazarded; he who bets against it, ought to hazard 5 shillings, according to what has been already said: hence the probabilities of gaining are $\frac{3}{8}$ and $\frac{5}{8}$.

But the mathematical expectation of the first player will be the product of 5 shillings, which he hopes to gain, by the probability $\frac{3}{8}$ which he has of gaining, or 1·87 shillings. The mathematical expectation of the second player will be 3 shillings multiplied by $\frac{5}{8}$, or 1·87 shillings equally.

It is necessary, in all equal bets, that the mathematical expectations of the 2 players be equal, as in the foregoing examples.

When the mathematical expectations are not equal, yet only differing by a small quantity, this difference will become very apparent after a great number of trials, for the favoured player will always gain, and the other lose. Thus, as we shall soon find, the mathematical expectation in lotteries is much stronger for the government, than for those who play; hence the government obtains a profit, with certainty, from the great number of chances which it has.

QUESTIONS TO LESSON VIII.

When the probabilities of gaining are the same, how are bets to be made?

When the probabilities of gaining are not the same, how ought the players to act?

What is mathematical expectation?

How ought mathematical expectancy to be regulated, in an equal bet?

When mathematical expectancy is not equal, what is the position of the players?

LESSON IX.

ON MORAL EXPECTATION, OR HOPE.

In estimating the mathematical probabilities of players, it has been rigidly done, without any regard to their relative positions. We ·have regulated the stake similar to a tribunal of justice; but as counsellors, as friends, we ought to consider the same game under another aspect, and as regulated according to the strictest laws of equity.

It is of moment to reflect that the same sum acquires more importance, when lost, than when gained. Suppose that a friend hazards a thousand shillings against a thousand shillings, in a game where the chances are on both sides equal.

According to what we have already stated, these conditions will be perfectly equitable; but if this friend thus hazards one-half of his annual income, and which is necessary to his sustentation, we have some right to ask, if the privations which he will have to impose on himself, in the event of losing,

E 2

should not be placed in the balance against the advantages, which he may reap in the event of gaining; we should represent to him that what he hazards should be far more valuable to one in his position, than that which he may gain.

The sum of one thousand shillings may, perhaps, be of far less importance to the other player, who may have large possessions, and who will, therefore, be little affected in case of a loss.

Thus we find, that, although the conditions may be equitable, the positions of the players are, notwithstanding, very different in this case.

The importance of the sum ought to depend on the property which we possess. In fact, one shilling is of greater value to a person who commands but 100 shillings, than to another who possesses 100,000.

The importance of a sum is estimated, by dividing that sum by the property which the person possesses, who hazards it.

The fraction thus obtained, is denominated the *moral value* of the sum; a friend who may have but 2,000 shillings, and who hazards 1,000, will hazard ½ of his property; that is to say, a sum as important to him, as 50,000 shillings would be to a person who possessed 100,000.

By calculating, in this manner, the importance of a sum, a more just idea may be obtained of the position of a player, always supposing that the principles of equality be observed, in the mathematical probabilities.

Let us suppose a player to possess 2000 shillings, and that he hazards 1000 against 1000 other shillings, the chances are therefore equal on both sides: he who hazards ought, according to what has been said, to consider that his risk will be represented by $\frac{1}{2}$; but if he gain, he will have 3000 shillings, and consequently his gain will be only represented by $\frac{1}{3}$. What he hazards, and what he expects to obtain will be thus found represented by the fractions $\frac{1}{2}$ and $\frac{1}{3}$: the difference of these fractions, or $\frac{1}{6}$, is the diminution of moral value that this game will cause to his property.

In making the calculation, we have,

$\frac{1}{2}$ of 2000 shillings=1000, sum hazarded.

$\frac{1}{3}$ of do.   do.   =666·66, sum looked for.

$\frac{1}{6}$ do.   do.   =333·33 diminution.

Again, suppose, as in the preceding example, that the second player possesses 100,000 shillings, the importance of the loss and gain to this player will be represented by the fractions

$\frac{1000}{100,000}$ and $\frac{1000}{101,000}$ or $\frac{1}{100}$ and $\frac{1}{101}$:

and the diminution of the moral value of his property will be the difference of these two fractions, or $\frac{1}{100,000}$ of 100,000=9·9 shillings. The *positions* of two such players are therefore very different.

It is easy to see by the preceding calculations

that all games, be they what they may, even when played in the most equitable manner, ought to produce a diminution in the moral value of property. This diminution may become, it is true, almost insensible, when only small sums are hazarded, relatively to the property possessed. Prudence should therefore place us on our guard against games which present themselves even under the most equitable form : this rule, which common sense indicates, is here established by calculation.

Some difficulty presents itself in the manner of appreciating the importance of a sum to an individual who actually possesses nothing. But as has been observed, the property possessed by an individual is at all events represented by his capability of employing his strength and industry, and which can be annihilated only with his life. It is only the individual dying of hunger who possesses absolutely nothing. " He who is enabled to procure by begging a sum of 10 pieces of gold annually," says Bernouilli, " will not accept for 50 sous the condition of renouncing this mode of gaining his livelihood for that of any other."

It is thus also with those who live by borrowing. Would they deny themselves for ever this resource for a sum even more considerable than that which would be required to liberate them from their debts ? If then the mendicant and the borrower will not renounce their species of trade, the first for a capital at least of 10 pieces of gold, and the second for one of 100, we consider the one as worth 10, and the other as worth 100 pieces; although, in ordinary language, one is said to be worth nothing, and the other less than nothing.

Moral hope or expectancy is then the product of the moral value of a sum by the probability there is of obtaining it.

Thus, in the example of the two players before cited, the probability of gaining being for both equal to $\frac{1}{2}$; and the moral value of the sums hoped for, being $\frac{1000}{5000}$ for the first, and $\frac{1000}{101,000}$ for the second player, the moral hope will be,

$\frac{1000}{6000}$ of his property to the first player,

$\frac{1000}{202,000}$ .... to the second—

or $\frac{1}{6}$ of 2000 shillings $=333{\cdot}33$ to the 1st;

$\frac{1}{202}$ of 100,000 do. $=495{\cdot}05$ to the 2d.[*]

Calculating the mathematical hope or expectancy in place of the moral hope, the result would be, 500 shillings for each player. From the preceding results, it is obvious that both players have some disadvantage in hazarding their money; but

* May not similar calculations and similar considerations direct the legislature in apportioning pecuniary punishments to crime? A nobleman who violates the laws of sobriety and order is fined 40 shillings, no more than what is required from the poor artisan or labourer under similar circumstances, though the property of the one may be 500 times that of the other.

the disadvantage is much more considerable for the first than for the second, who possesses the larger property.

QUESTIONS TO LESSON IX.

Is it sufficient in playing to consider only the mathematical expectation ?

Has a sum the same importance whether gained or lost ?

On what will depend the importance of a sum ?

How do you estimate the importance of a sum ?

What is the *moral value* of a sum ?

How do you calculate the diminution of the moral value, that the game, even the most equitable, causes to a fortune ?

Can an equitable game of any sort prove advantageous to the players ?

What does prudence direct with regard to games of hazard ?

Is there not a difficulty experienced in appreciating the importance of a sum ?

What is moral expectation ?

Give an example in calculating moral expectancy.

LESSON X.

LOTTERIES*.

THE advantage to him who keeps a lottery consists in this, that his mathematical hope or probability is generally considerably stronger than that of the players ; and that he secures this advantage by the great number of drawings or trials which

* See Note I.

occur. We shall endeavour to make this understood by an examination of the Genoese lottery*.

The Geneva lottery is composed of 90 numbers, 5 of which issue at each drawing; thus, in *simple drawing*, a sum is hazarded on a certain named number, and if this number be one of those drawn, 15 times the value of the stake is gained. But let us see whether the mathematical hope of the player be equal to the value of the sum which he hazards. The lottery is composed of 90 numbers : there are in all 90 chances, of which 5 are favourable to the player. The probability that the number of the ticket held by the speculator will be one of the five to be drawn, is represented then by the fraction $\frac{5}{90}$ or $\frac{1}{18}$. If the stake be 100 shillings, in order that the game be equitable, the gain required to reimburse the speculator should be 18 times the amount of the stake, and not 15 times— 1800 shillings, and not 1500. What is then the position of the speculator?—The same as that of one who, playing with another on supposed equal terms, finds himself robbed by him each time that he wins, of a certain portion of his gain, without power on his part to claim a similar share. It is plain how disadvantageous must be the position of the player. Supposing that an artisan, who had imprudently purchased a ticket for 100 shil-

* See Note II.

lings, and fearful of the consequences to which the drawing might expose him, wished to dispose of it on equitable conditions, what is the value which he has a right to expect? We find that for 18 shillings he cannot obtain more than 15. Thus, in place of 100 shillings, he can only hope for $\frac{15}{18}$ of 100 or 83s. 4d., being the mathematical hope ; he must therefore experience a loss of 16s. 8d., which may be considered as a species of tax paid for the pleasure of playing. We may add, that we have not taken into calculation the diminution of the moral value which the property of the player is subjected to in the speculation. To give an example of the calculation which should be made on the preceding hypothesis—suppose that this same artisan who risked 100 shillings, possessed altogether but 1000, we have for the moral value of the sum risked

$$\frac{100}{1000} = \frac{1}{10},$$

and for that of the sum hoped for, say

$$\frac{83\cdot33}{1083\cdot33} = \frac{1}{13} \text{ nearly.}$$

The diminution of the moral value of the fortune of the artisan, will be then $\frac{1}{10}, - \frac{1}{13},$ or $\frac{3}{130}$; that is, about 23 shillings. Thus, this artisan, who paid 100 shillings for a ticket which was worth but 83s. 4d., suffers still, in the moral value of his property, a diminution of 23 shillings.

It may be asked, certainly, if there be not some other modes of speculating in the lottery. Such do exist, but they are yet more disadvantageous; we proceed to exhibit them.

Determinate Drawing.—A certain sum is placed on a number, and moreover the order is declared in which it shall *issue.* Here there is *but one* chance out of 90. Strict justice would entitle the speculator to 90 times his stake; he receives but 70.

In hazarding 10 shillings, he ought, in the event of winning, to receive 900; he receives but 700: the other 200 become the property of the banker.

Ambe.—A certain sum is hazarded on two named numbers, and in the event of winning 270 times, the stake is received. Now calculation shows that 90 numbers can be arranged two and two in 8010 different ways: there are then 8010 chances, and but 20 in favour of the speculator— because with the five numbers drawn, there can be but 20 different arrangements made of two numbers. The probability of success is then $\frac{19}{8010} = \frac{1}{400.5}$. For one shilling therefore, which may be hazarded, $400\frac{1}{2}$ should be received; 270 only are given.

Ambe determinée.—Two numbers are selected, and the order of their issue determined. There is here but one chance out of 8010. In the event of winning, 8010 times the stake ought to be received; 5100 only are given—the sum retained being nearly $\frac{3}{8}$ of that returned to the speculator.

Terne.—This is formed by the issuing of three named numbers. Calculation shows that with 90 numbers, 704,880 arrangements of three numbers can be made; and with the 5 numbers drawn, there can only be 60. The speculator has then 60 chances out of 704,880, the probability of winning being $\frac{1}{11,748}$, he ought to receive 11,748 times his stake should he succeed—5500 times only are given ; thus, in the event of winning, he divides his profit with the treasurer, who takes more than one half.

Quaterne.—This is formed by the issuing of 4 named numbers. Calculation shews that with 90 numbers, 61,334,560 different arrangements of 4 numbers can be formed, and with the 5 numbers which supply the drawing, there can only be formed 120. Out of 61,324,560, the speculator has but 120 chances, and the probability of winning will be $\frac{1}{511,038}$. Should the speculator succeed, he ought to receive 511,038 times his stake ; he *really* receives only 75,000 times. It is obvious, therefore, how unfavourable the position of the speculator must be.

Setting aside the consideration of the injury inevitably produced by gaming, the individual is placed in the position of a person who, when playing with another, would give no indemnification in the event of his losing, and who, in the event of winning, deducts always about $\frac{6}{7}$ of the profit.

Well may we exclaim against such an iniquity !
Such, is, however, the position of the speculator
who hazards on a *quaterne ;* ignorance alone can
excuse it.

Neither *Terne déterminé* or *quaterne déterminé*
is adopted.

Formerly *Quine* was used, or the issuing of 5
named numbers. The treasury, in the event of
gaining, deducts then more than $\frac{44}{43}$ of the profit
of the speculator.

Although the Geneva lottery be defended in this
country (Netherlands), it will be satisfactory to
know what the true value of a lottery ticket of 100
francs* is. The following Table exhibits the several
values, independent of the diminished moral value
which the property of the player or speculator ex-
periences :—

TABLE, No. 8.

	Sum hazarded.	Value of the sum hazarded.
Extrait	100	...... 83·33
Extrait déterminé	100	...... 77·77
Ambe	100	...... 67 50
Ambe déterminé	100	...... 63·67
Terne	100	...... 46·82
Quaterne	100	...... 14·68

The third column indicates the mathematical
expectation of the speculator, in other words, the

* A franc is equal in value to ten-pence.

F

several sums for which a lottery ticket of 100 francs will actually sell for.

The greater the number of chances, (other circumstances being the same,) the greater becomes the difficulty of seeing the agreement between the result of calculation and experience as we have found; this will explain easily why the treasury, which seeks to secure these advantages, obtains so much greater profits from the events that depend on the greater number of chances.

The profits derived by the government from a lottery institution, is a sort of tax on which it reckons with as much certainty as on that of any other kind : as a proof, we have only to cast our eyes on the following Table, which shows the sums that the Paris lottery have put into circulation during 5 years*.

TABLE, No. 9.

| Years. | Hazarded. | SUMS | |
		Received by the Speculators.	Returned to the treasury.
	Francs.	Francs.	Francs.
1816....	19,552,000	13,383,000	6,169,000
1817....	21,461,000	16,513,000	4,948,000
1818....	29,371,000	22,765,000	6,606,000
1819....	27,524,000	22,306,000	5,218,000
1820....	29,036,000	19,783,000	9,253,000
Total ..	126,944,000	94,750,000	32,194,000
Mean ..	25,388,000	19,950,000	6,438,000

* Statistics of Paris.

The treasury received, then, a little more than the fourth of the sum hazarded.

M. Quetelet then enters on the subject of a contemplated lottery in the Netherlands, where he shews that the speculators are subjected to a loss of 22 per cent., according to the mathematical probability.

QUESTIONS TO LESSON X.

What advantage has the banker in the game of lottery?
What is the Genoese lottery?
What is simple drawing?
How much per cent. is lost, by embarking in simple drawing?

Give an example of the loss of an individual, though fortunate in playing, at simple drawing.

What is determinate drawing?
What is ambe?
What is determinate ambe?

What is the expectancy of a player who hazards a sum on a *terne*?

Exhibit the great disadvantage of the player who embarks in a quaterne.

What is the loss experienced on 100, when embarked in simple drawing, ambe, terne, &c.?

Why is the benefit to the treasury more considerable in the quaterne scheme, than in that of simple drawing, or ambe?

Ought a government to reckon on the gains from a lottery?

What is the mean gain from the Paris lottery?

LESSON XI.

ON THE CALCULATION OF A PROBABILITY, WHEN
THE NUMBER OF FAVOURABLE CHANCES ARE
NOT KNOWN.

WE have hitherto supposed, that all the chances,
both favourable and unfavourable, of an event the
probabilities of which we seek, can be estimated.
We shall now consider cases where these circumstances do not obtain. We suppose only that the
total number of chances are known.

Suppose an urn to contain 2 balls, the colours of
which are unknown. One of the balls, a white one,
is drawn which is replaced for the purpose of pro-
ceeding with the second drawing; but the second
time you draw a white ball, and the question arises
as to the colour of the balls. Here there can be
only two hypotheses : both balls are white, or
only one, the other being of some other colour,
red for instance. These two hypotheses are only
probable, and it is required to determine their relative probabilities. Theory offers the following
method. The probability of the event occurring,
is calculated on the different hypotheses that can
be formed, and the numbers obtained are proportional to the probabilities of these hypotheses.
Thus, in the first hypothesis, the two balls being
white, there is a certainty, that at each drawing a

white ball will be taken, the number sought for is,
therefore, 1. In the second hypothesis, one of the
two balls being white, there is for the probability
of its being drawn $\frac{1}{2}$, and for both being drawn
successively $\frac{1}{2} \times \frac{1}{2} = \frac{1}{4}$.

Thus the numbers 1 and $\frac{1}{4}$ being obtained, enable us to calculate the probability of the expected
event, by the two only hypotheses which can be
formed; but these numbers may be regarded as
proportional to the probabilities of the two hypotheses; consequently, the probability of the hypothesis of 2 white balls, is to that of 1 only, as 1
to $\frac{1}{4}$, or as 4 to 1.

The foregoing principle may be thus stated:—

*The probabilities of hypotheses (as the causes
of events) are proportional to the probabilities
which these hypotheses afford for the observed
events.*

To sustain, in a wager, the first hypothesis, it
must be considered that you have 4 chances out
of 5; in other words, the probabilities of the 2
hypotheses are $\frac{4}{5}$ and $\frac{1}{5}$. Retaining the foregoing
example, let us suppose the probability were re-
quired, of once again obtaining a white ball at a
third drawing. It will be necessary to consider
the two former hypotheses as two urns, from one
of which you ought to extract a white ball: the
problem returns then to a compound probability.
The calculation will be as follows:—The pro-

bability of the first hypothesis is $\frac{4}{5}$, which must be multiplied by the probability of the event, in this hypothesis ; but, we have here the certainty of drawing a white ball ; thus it will be necessary to take $\frac{4}{5}$ of 1, or $\frac{4}{5}$. The probability of the second hypothesis is $\frac{1}{5}$; and on this hypothesis the probability of drawing a white ball is $\frac{1}{2}$; the product of these numbers is $\frac{1}{10}$. The probability of drawing a white ball, a third time, will be then $\frac{4}{5}$ + $\frac{1}{10}$ or $\frac{9}{10}$.

Reduced to a principle, it may be thus stated: *The probability of a new simple event is obtained by calculating, according to the past events, the probabilities of the various possible hypotheses, and making the sum of the products of these probabilities taken in each hypothesis equal to the probability of the event.*

QUESTIONS TO LESSON XI.

What is the object of this lesson ?

How do you calculate the probabilities of the causes of events ?

What is the general rule?

Apply the rule to an example.

How do you calculate the probability of a new event, according to past events ?

What is the general rule ?

LESSON XII.

ON THE CALCULATION OF A PROBABILITY, WHEN THE NUMBER OF CHANCES IS UNKNOWN.

WE suppose, in what follows, that neither the number of favourable chances of an event, or the total number of chances, are known; the results of many experiments only are known, and it is required to calculate the probability of the event by their means.

The resolution of this question, so interesting to all the sciences founded on observation, applies itself to calculations of a superior order.

Fortunately the results of these calculations are easy of comprehension.

Let us begin by examining the case where an event occurs any number of times in succession.

Theory indicates that the probability of this event being reproduced once again, in succession, is equal to the number augmented by unity, divided by the same number, augmented by two units.

Example.—Eleven planets have been already discovered, all of which move round the sun in the same path : it is required to know the probability, should another be discovered, that it would be also found to move in a similar manner to them. According to the rule, we divide $11 + 1$ by $11 + 2$, and we have for the probability required $\frac{12}{13}$.

It is required to know the probability of the sun re-appearing to-morrow on the horizon.

If the calculation be made only from the number of times which we have known the phenomenon to recur, deducting other considerations, the calculation will be made as follows :—To the 1st of January, 1838, we have 5,842 years; or, 2,132,330 successive returns of the sun, since the creation, that period being fixed at 4,004 years before the Christian era; the probability of another return of the sun on the horizon at this epoch will be $\frac{2,132,331}{2,132,332}$, the odds of 2,132,331 may be taken against 1, that the phenomenon will take place.

Example.—You have had an opportunity of verifying 20 assertions of a person, and you have been satisfied as to their truth; let it be required to know what the probability is, that the 21st assertion will be equally true. By dividing 21 by 22 you obtain the probability required, viz. $\frac{21}{22}$. Thus there are 21 chances against 1, that a new assertion, made by this same person, will be true, similar to the former ones. We shall now see, that after having observed the same event many times in succession, there are probabilities, more or less strong, to believe in the return of that event. One is disposed to think that a cause exists which facilitates its reproduction; but theory offers a very simple method, by which to calculate the probability that such cause really does exist.

Probability is a fraction which has for its denominator the number 2, multiplied as many times into itself, as the event has been observed consecutively; and, for its numerator, this same product, less 1.

Example.—What is the probability that a far greater facility exists for the movement of the planets around the sun, in one direction, than in another.

Eleven planets have been observed to move in the same path, the probability that a greater facility exists for their movement in that path, is then $\dfrac{2^{12} - 1}{2^{12}}$ or $\dfrac{4095}{4096}$. Thus, there are 4,095 chances to 1 that the probability of the event, constantly observed, is superior to $\frac{1}{2}$.

The probability that a cause exists for the periodical return of the sun on the horizon, is a fraction which approaches so nearly to unity, that it may be considered as certainty. The number of chances against 1, amounts to more than 64 billions of figures, and is incapable of being given in ordinary language.

The probability may be thus calculated, that an event will be reproduced many times, which had already been observed for a certain number of times in succession.

This probability will be represented by a fraction which has for its numerator the number of observations made plus 1, *and for its denominator the same number plus* 1, *and plus, also, the number of times that the event ought to be reproduced.*

Example.—What is the probability, that, if 3 more planets be discovered, they will move in the same path as the 11 others already known? Here, then, we divide 11 + 1, or 12 by 11 + 1 + 3, or 15; and we have for the required probability $\frac{12}{15}$ or $\frac{4}{5}$. For the probability that the sun will appear 10 times on the horizon, we have for the 1st of January, 1838, $\frac{2.192,331}{2,132,341}$. We see that the probability goes on continually decreasing, as the number of expected returns become greater.

The preceding rules are of great value in the sciences of observation. For example, 4 experiments afford similar results; and it is thence deduced, conformably to what has been stated, that the probability of a similar result, from a fifth experiment, is $\frac{5}{6}$. While the probability that a cause exists, which favours the recurrences of the observed results, is $\frac{31}{33}$.

We have hitherto supposed, that only one sort of events have been observed, which is always being reproduced; in what follows, we shall admit two sorts of events, each of which is only produced a certain number of times.

Example.—From an urn 17 white balls, and 3 black, had been drawn in 20 consecutive trials; and it is required to know the probability of drawing, at the 21st trial, a white ball. In this case, it is necessary to divide 17 plus 1, or 18, by the number 20 plus 2, or 22, and the probability of drawing again a white ball, at the 21st trial, becomes $\frac{18}{22}$. In the same manner, the probability of drawing a black ball becomes $\frac{4}{22}$, and these 2 probabilities added together are equal to unity ; the urn is supposed, here, to contain only black and white balls.

Of 116 comets, the orbits of which have been calculated in the Astronomical work of Delambre, 23 only have their perihelion* greater than the distance of the earth from the sun. The probability that a new comet ought to be arranged in the same class, will be 23 plus 1, or 24, divided by 116 plus 2, or 118; that. is, the fraction $\frac{24}{118}$, and the contrary probability is $\frac{94}{118}$. There are, then, 24 against 94, as the odds in a wager, in favour of the looked for event, or about 1 to 4.

In reducing what has been stated to a rule, we find, that *when we observe two sorts of events, the probability that one of these events will reproduce itself once, is a fraction which has for its*

* The perihelion distance of a star, is its nearest distance from the sun.

numerator, the number of times that the event in question has been observed plus 1; and, for its denominator, the total number of observations plus 2.

The calculations become somewhat complicated when we desire to determine, as in the present case, the compound probabilities. Fortunately it is found that the results are nearly the same as those which are obtained, when considering the favourable and unfavourable chances as being numerically in the same ratio as the observed events.

Example.—According to the preceding calculation, we have seen that $\frac{24}{118}$ is the probability that a newly observed comet will have its perihelion distance greater than the distance of the earth from the sun, and the probability against it is $\frac{94}{118}$.

Considering 23 as being the number of chances favourable to the event, and 93 as the number opposed; this allows in all but 116 possible chances, we have for the probabilities required $\frac{23}{116}$ and $\frac{93}{116}$. Thus the fractions estimated in this manner differ little from those which we had previously obtained.

The probability that two comets shall be observed in succession, the one having its perihelion distance less than the distance of the earth from the sun, and the other having, on the contrary, its perihelion distance greater, is a compound probability formed by the product of the two fractions $\frac{23}{116}$ and $\frac{93}{116}$.

The greater the number of observations, the less will be the liability to error in calculating by the last mode, which supposes the total number of chances to be known, and the number of favourable chances.

These ideas will be developed in the following lesson.

QUESTIONS TO LESSON XII.

What is the object of this lesson?

How do you calculate the probability that an event which had been observed many times following, will occur once again?

Apply the rule to different examples.

How do you calculate the probability that there exists a cause which facilitates the reproduction of an event which had been observed many times following?

Show the application of the rule to different examples.

How do you calculate the probability that an event, which had been observed many times following, will again occur a given number of times?

Apply the principles to the science of observation.

When two descriptions of events have been observed, how do you calculate the probability that one of these events will be produced?

Can you simplify the calculations in the preceding examples?

LESSON XIII.

THE MODE OF TAKING MEAN RESULTS.

IT is often found necessary to take a mean value of many numbers : all the numbers are added together, and the sum is divided by their quantity. The result is the mean value.

Example.—If it be required to determine the duration of human life at a certain epoch, and in a given country, the age attained by each one of a great number of men under the most diversified circumstances, must, with the deaths, be marked; the sum of these ages divided by the number of deaths, is the mean duration of life.

It is obvious that the mean value is known with so much the greater precision, as the number of observations brought to the inquiry are multiplied. It is also plain that these must not be limited to certain professions or conditions ; but all must be indiscriminately admitted, so that from the multitude and the promiscuousness of the elements, the accidental variations may compensate one another, and thus a mean general result be formed.

We have, in fact, seen that in consequence of a vast number of experiments, the multiplicity of the chances cause what is accidental and fortuitous to disappear, and that uniform effects only remain from constant causes ; in short, that there is no hazard in natural facts considered in a great number.

A competent knowledge may be acquired of the precision of a mean result, by the following mode: —It is sufficient, for example, to divide into two parts the whole of the observed values, the number of which is supposed to be very great, and to take for each of the parts the value of the mean result; for if these two values differ extremely little the one from the other, each of them may be considered as accurate. Nothing is better adapted than this species of proof to afford evidence as to the exactness of statistical results ; and it is almost useless to offer consequences to the student which are not verified by comparisons of the mean values.

Example. — Let it be supposed that an urn contains an unknown number of black and white balls, the unknown ratio of these two numbers may be determined. To effect this, it will be necessary to make a great number of trials, by extracting one ball from the proposed urn, and after having noted the colour, returning it. The number of times that a white ball appears, to be reckoned, also the number of the black. The ratio of these two numbers, which may be designated by m and n, may at first differ considerably from the unknown numbers M and N;

but the variable quotient $\dfrac{m}{n}$ will continually ap-

proximate to the fixed quotient $\dfrac{M}{N}$. Thus, supposing that the number of trials made be very
great, and that m and n represent the numbers
respectively of white or black balls which had
been drawn from the urn, the ratio $\dfrac{m}{n}$ will differ

extremely little from the ratio of $\dfrac{M}{N}$. The differ-

ence $\dfrac{M}{N} - \dfrac{m}{n}$ may be positive or negative, that
being fortuitous; but the actual value of this difference will necessarily be an extremely small
decimal fraction. Supposing now that, after
having made the number of trials, which may
be represented by r, the operation be repeated,
and let the number of trials in this second
operation be r or another larger number r'.
The ratio $\dfrac{m'}{n'}$ of the respective numbers of white
and black balls issuing during this second operation, will differ also extremely little from the fixed

ratio $\dfrac{M}{N}$: thus the quantities which $\dfrac{m}{n}$ and $\dfrac{m'}{n''}$

differ from one another, and differ from $\dfrac{M}{N}$ dimi-
nish indefinitely and without limit as the numbers
r and r' augment; that is to say, that the numbers

r and r' of the trials may be made so great that there will be no appreciable difference between the ratios deduced from the one and the other operation.

One of the most simple means of verifying the numbers which furnish the multiplied observations consists, as we have said, in dividing at hazard, the series of these observations into many parts, and in comparing the values which are separately deduced from each of the parts.

The use of these rules evidently supposes that the composition of the urn does not change during the whole time of the trials.

These rules may, without doubt, be applied to cases where changes arise in the nature of the causes, and the effect of these changes may even be known; but in this case it will be necessary to consider separately the intervals in which the cause remains constant, and to multiply the observations relating to each of these intervals.

The most common sources of error and uncertainty as to the results which many writers have adduced from statistical researches are, 1st, the want of correctness in the primary observations collected by various means not capable of being compared; 2d, the small number of these observations, which do not allow of division into series, and of obtaining a separate result for each; 3d, the alteration, whether uniform or irregular, that the causes are subject to, during the duration of the observations.

How do you find the mean value of many numbers?

What is it that tends to increase the exactness of a mean value?

How do you recognise the correctness more or less great of a mean value? Give an example.

Will mean values enable you to discover the changes which arise in the nature of the causes of events?

What are the most common sources of error and uncertainty which affect observations?

LESSON XIV.

ON THE MEASURE OF THE DEGREE OF APPROXIMA-
TION OF A MEAN RESULT, OR THE RULE OF THE
LESSER SQUARE.

IN taking a mean result, the degree of approximation does not depend alone on the number of quantities added together, it depends yet more or less on the diversity of these quantities It is expedient to form exact ideas on this degree of approximation, and to show that the precision of the result is a measurable quantity which may be always expressed in numbers.

We shall first enunciate the rule which ought

to be followed in order to find this numerical measurement with precision.

The mean value of all the particular values which are to be considered must be first found. The square of each of these particular values to be then taken, and a second mean value sought, namely, that of the squares. In this manner there will be obtained for the two means, two numbers, the first of which we shall call A, and the second B. Subtracting from B the square of the number A, divide double the remainder by the number of the particular values which are under consideration. By extracting the square root of the quotient a quantity will be found which we may call g, and which will serve to measure the degree of approximation; the less the value of g, the nearer the calculated mean A, approximates to the exact value sought.

Example.—Suppose there be found 4,000 particular values, namely:—

> 1,000 equal to 2,
> 2,000 equal to 5,
> and 1,000 equal to 12.

In general the observed quantities are all unequal, and they cannot be reduced, as those above have been, to a small number of different values; but the present object is only to shew the mode of calculation.

The sum of the observed values is $(1,000 \times 2)$

$+ (2,000 \times 5) + (1,000 \times 12)$ or 24,000, and this sum divided by 4,000, the number of the quantities, gives 6 for the mean value which we have represented by A. The sum of the squares of the values is $1,000 \times 4 + 2,000 \times 25 + 1,000 \times 144$ or 198,000; dividing this sum by 4,000 we have 49½ for the mean of the squares which we have represented by B. From this we take the square 36, or the square of the mean value A, the remainder is $\frac{27}{2}$. We double this remainder and divide 27 by 4,000; lastly, we extract the square root of the quotient $\dfrac{27}{4000}$ or $\dfrac{1080}{160,000}$; this root is $\dfrac{1}{400}\sqrt{1080}$; by completing the operation, we have 0·08216, or very nearly the 82-thousandth part*. It is this fraction which shews the degree of approximation of the mean result.

The respective precisions of two results are therefore in the inverse ratio of the fractions obtained in this manner.

The mean result then of an infinite number of observations is a fixed quantity, subject to no contingency, and which has a certain ratio with the nature of the facts observed.

Each of the particular values can be compared

---

* This operation may be more readily accomplished by logarithms.

with it, and the difference between this particular
value and the fixed one is called the error or the
deviation, which will be the mean result of an infinite number of observations.

These errors have probable limits, that is to
say, it is extremely likely that the error committed,
greater or less, will not exceed a certain quantity.

Other limits exist, much nearer, for which the
probability of error is only $\frac{1}{2}$; so that it may occur one way or the other, or the error may exceed
these limits, or it may be comprised in it.

Generally, to determine the mean result of a
great number of particular values, we should measure with an instrument, the precision of which
may be augmented as much as we please, by augmenting more and more the number of the values
observed.

It remains now to determine the probability
which exists that the fixed mean sought is comprised between the limits proposed, $A + D$ and
$A - D$: A is the mean result found, and D is a
proposed quantity which is added to the value A
or subtracted from it. The following table gives
the probability P of error positive or negative
greater than D; and this quantity D is the product of g, of which we have already spoken, by a
proposed factor d.

TABLE, No. 10.

d		P
0·47703	....	$\frac{1}{2}$
1·38591	....	$\frac{1}{20}$
1·98495	....	$\frac{1}{200}$
2·46130	....	$\frac{1}{2000}$
2·86783	....	$\frac{1}{20000}$

Each of the numbers of the column P gives the
existing probability that the exact value of the
fixed mean, (which it is the object to discover,) is
comprised between the limits $A+D$ and $A-D$.
The quantity D is equal to the product $g \times d$, as
we have stated. It will be seen by the Table that
the probability of an error, greater than the product of g, by 0·47708 is $\frac{1}{2}$. The chance is then
as 1 to 1 that the error committed will not surpass the product of g by 0·47708. In the former
calculation for determining the precision of the
mean value of a series of numbers we have found
for g the quantity 0·08216 ; multiplying this
quantity by 0·47708, we obtain 0·039 nearly.
Thus the chance is as 1 to 1 that the error committed will not exceed 0·039.

The probability of an error greater than the
product of g by 1·38591 is considerably less than
the preceding ; being but $\frac{1}{20}$. The chances are
19 to 20 that the error of the mean result will not
exceed this second product.

The probability of an error greater than the

preceding becomes extremely small as the factor
D augments. It is not more than $\frac{1}{2000}$ when d
approaches to 2. The probability falls afterwards
below the $\frac{1}{2000}$. Lastly, the chances are considerably more than 2,000 to 1 that the error of
the mean result will be below the triple value found
for g. Thus in the example already taken, where 6
is the mean result, it may be considered as nearly
certain that this value 6 does not err by a quantity
triple the fraction 0·08216 that the rule has given
for the value of g. The fixed mean sought for is
then comprised between $6 - 0·246$ and $6 + 0·246$.

In order to facilitate the calculation of the value
of g, the particular values may be considered as
equal amongst themselves as they differ but little,
and by attributing thus a common magnitude to a
certain number of particular values, the calculation will be rendered much easier.

The result of this manner of calculating the
precision of a mean value is, that the precision
augments as the square roots of the number of observations. Suppose, for example, that in one
case 100 observations have been made, and in another 400 observations, the precision of the mean
deduced from the first series of observations will
be to that deduced from the second, all the circumstances being in other respects equal, as 10
to 20, that is to say as the square roots of the
numbers 100 to 400. Hence, for the same in-

quiry, the precision of the mean result changes in proportion as the number of the observed values augment. It is double if the number of the values becomes four times greater; treble, if this number becomes nine times greater, and so on. This consequence is simple and remarkable; it ought to be known to all those who undertake statistical inquiries; it shews how necessary it is to multiply observations in order that the result shall acquire a given degree of exactness.

QUESTIONS TO LESSON XIV.

When the mean value of many quantities is obtained, can you measure the precision of the result?

State the rule derived from this operation.

Apply the rule to an example.

Can you find the probability that the mean sought for will not pass certain assigned limits?

Give examples in calculating this probability.

How do you simplify the calculations?

In what ratio does the precision of the mean value augment relatively to the number of observations?

Give a numerical example.

LESSON XV.

APPLICATION OF CALCULATION TO THE
PROBABILITIES OF HUMAN LIFE.

ONE of the most interesting applications of the calculations of probabilities is, the formation of tables of mortality, their object being to make known the law according to which a certain number of individuals, born at the same period, successively perish.

Halley, who constructed the first table of mortality (in 1693), employed the following method: he made, for the city of Breslaw, in Silesia, an enumeration of all individuals who, in the period of four years, died between 0 and 1 year, between 1 and 2 years, between 2 and 3 years, and so on to the most advanced period of life; at the same time considering the population as stationary, or as affording annually a number of deaths equal to the number of births, and that all the individuals whose deaths he enumerated had been born at the same time, he deduced from their respective ages, the law according to which they successively perished.

He took then the sum of all the deaths, deducting from it the number of infants which died between 0 and 1 year, the remainder indicated the number of survivors; from this last remainder he

H

deducted the number of infants which died between 1 and 2 years, to obtain the number of survivors, and so on.

TABLE, No. 11.

Dr. HALLEY'S TABLE

ON THE BILLS OF MORTALITY AT BRESLAW.

Ages.	Persons living.	Age	Persons living.	Ages.	Persons living.	Ages.	Persons living.
1	1000	24	573	47	377	70	142
2	855	25	567	48	367	71	131
3	798	26	560	49	357	72	120
4	760	27	553	50	346	73	109
5	732	28	546	51	335	74	98
6	710	29	539	52	324	75	88
7	692	30	531	53	313	76	78
8	680	31	522	54	302	77	68
9	670	32	515	55	292	78	58
10	661	33	507	56	282	79	49
11	653	34	499	57	272	80	41
12	646	35	490	58	262	81	34
13	640	36	481	59	252	82	28
14	634	37	472	60	242	83	23
15	628	38	463	61	232	84	19
16	622	39	454	62	222	85	15
17	616	40	445	63	212	86	11
18	610	41	436	64	202	87	8
19	604	42	427	65	192	88	5
20	598	43	417	66	182	89	3
21	592	44	407	67	172	90	1
22	586	45	397	68	162		
23	579	46	387	69	152		

The method which we are about to exhibit, supposes a population stationary, which seldom occurs: if, however, it leaves something to be desired, on the side of accuracy, it presents great advantages in the simplicity of its application.

The following Table*, which gives the law of mortality in the southern provinces of the Netherlands, shows how 100,000 individuals, born at the same time, successively perish.

LAW OF MORTALITY.

Years.	Individuals.	Years.	Individuals.	Years.	Individuals.	Years.	Individuals.
0	100.000	28	45,866	56	27,155	84	2,929
1	77,507	29	45,284	57	26,357	85	2,429
2	69,470	30	44,709	58	25,547	86	2,000
3	64,799	31	44,147	59	24,727	87	1,619
4	61,899	32	43,589	60	23,890	88	1,285
5	59,864	33	43,023	61	23,041	89	998
6	58,726	34	42,448	62	22,176	90	744
7	57,800	35	41,857	63	21,296	91	537
8	57,129	36	41,249	64	20,402	92	378
9	56,557	37	40,629	65	19,493	93	267
10	56,077	38	39,990	66	18,571	94	204
11	55,660	39	39,335	67	17,636	95	150
12	55,409	40	38,670	68	16,688	96	105
13	54,919	41	37,999	69	15,731	97	76
14	54,569	42	37,322	70	14,761	98	54
15	54,226	43	36,638	71	13,769	99	38
16	53,883	44	35,948	72	12,781	100	25
17	53,533	45	35,252	73	11,718	101	19
18	53,167	46	34,549	74	10,697	102	16
19	52,643	47	33,840	75	9,679	103	13
20	51,956	48	33,125	76	8,706	104	10
21	51,132	49	32,406	77	7,810	105	7
22	50,309	50	31,671	78	6,977	106	4
23	49,498	51	30,940	79	6,213	107	2
24	48,703	52	30,199	80	5,501	108	1
25	47,939	53	29,452	81	4,798	109	0
26	47,218	54	28,698	82	4,131		
27	46,528	55	27,871	83	3,504		

* I have added another, and more perfect Table in the Appendix, subsequently furnished by M. Quetelet in his work, " Sur l'homme," &c.   Note III.

H 2

We shall presently show the principal applications of this Table.

Probability of life may be considered as the number of years after which the probability of existing and of not existing, become the same; or rather, the number of years after which individuals of the same age are reduced one-half. The preceding Table shows, that of the 100,000 individuals which have been supposed to be born at the same time, but 500,000 remain between the 22nd and 23rd year: the probability, therefore, that a newly-born infant shall live to $22\frac{1}{2}$ years, viz. the probability of life, is $\frac{50,000}{100,000}$ or $\frac{1}{2}$. The probability of life in France, is $20\frac{1}{2}$ years nearly, according to the bureau of longitude; in England it is 27 to 28 years. The probability of life is very short in large towns; it falls at Paris to between the 8th and 9th year; at London a little before the 3rd year; at Vienna a little before the 2nd year; a little after at Berlin; and at Brussels after the 23rd year.

Mean life is calculated, by supposing that an equal division is made of all the ages of the individuals which have been considered in the Tables of Mortality.

TABLE, No. 12.

	Years.	Months.
It is in Switzerland	37	1
Southern provinces of the Netherlands	30	4

	Years.	Months.
France	28	9
Northampton	25	2
London	17	11
Berlin	17	1
Vienna	15	9

By means of the Table of Mortality, the probability of continuing to exist, for a certain number of years, may be determined to a certain age not specified. If it be required to know, for example, what the probability would be of living 12 years for an individual who had attained 30 years.

The number of survivors at 30 and 42 years, will be found in the Table respectively, 44709, and 37322. The first number ought to be considered as the total number of chances ; and the second as the number of favourable chances; the probability required will be then $\frac{37322}{44709}$; for France it will be found $\frac{355400}{438183}$; these probabilities are nearly equal.

If the probability of life be required at 30 years, one half the number of individuals of that age must be taken, which gives 22,354. This number corresponds in the Table to 62 years nearly; there is then 1 to 1 that the individual of 30 years shall attain 62 years; it may be said that the probability of life for a man of 30 years is 32 years. In France it is not quite 30 years.

The probability of life is greater or less according to the age: it is at its *maximum*, in nearly all countries, about the 4th and 5th year.

In the southern provinces of the Netherlands, the probability of life is at its maximum about the 5th year, and its value is from 47 to 48 years; in France, it falls between the 4th and 5th years, and is nearly 46 years.

The maximum of mean life falls between the 5th and 6th years in England, between the 6th and 7th years, according to Price; in France, between the 5th and 6th, according to Duvilard, and its value is from 41 to 45 years.

The probability may also be determined for the continued existence for a certain number of years, of two persons, their ages being given. This probability is then composed of two simple probabilities that each of these persons will live to the period named; for example, what is the probability that an individual 30 years of age, and his wife 20 years of age, shall continue to live 12 years. Multiplying the fraction $\frac{37322}{44709}$ by $\frac{43589}{51956}$, the product will express the probability of living 12 years longer; the latter fraction expressing the probability which the individual who had attained 20 years, has, of living still 12 years.

Thus may be calculated the probability that three, four, or any greater number of persons, may have of living for any time named.

The law of mortality may be rendered sensible

to the eye by a geometrical construction; (see the diagram ;) on a right line OA set off a certain number of equal parts which will represent the equal times or years. From each of the points of division raise perpendiculars to represent, by their lengths, the number respectively of the survivors at each period. The first perpendicular OB, will represent a certain number of individuals supposed to be born at the same time. The second will represent the number of survivors at the age of one year, and so on. Each perpendicular representing the number of survivors; and this perpendicular decreasing insensibly in proportion as the survivors become extinct up to the time when all cease to exist, which point represents the utmost duration of life.

The curves constructed in this manner for different countries, will not be exactly similar. These lines become shorter more or less rapidly according as the mortality is more or less remarkable.

In this construction, the second line indicates the progressive decrease in the number of individuals born at the same time, in the southern provinces (of the Netherlands). The upper line is applicable to Sweden, the fourth to Northampton, and the fifth to London.

The Table of Mortality which we have given, may also serve to determine, how far the individuals of a determinate age, may be calculated on in a population; this constitutes the law of population *; taking in effect the sum of all the numbers which the Table contains, and if this number be considered as representing the population, the particular numbers in the Table will represent the individuals of different ages of which the population is composed.

The calculation of probabilities has made known a singular fact, that more males than females are born; and this observation has been made in all countries. The ratio of male to female births, is—

TABLE, No. 13.

In the Netherlands............... as 1   to   0.9427

France 1   ....   0·9375

Kingdom of Naples.................. 1   ....   0·9560

England 1   ....   0·9470†

It is remarkable, from the observation of ten years, that the ratio is not the same for the towns and the country in the kingdom of the Netherlands. For the towns it is found as 1 to 0·9480, and for the country, 0·9375.

In estimating the population, it is also usual to find the ratio of the population to the births and deaths; these ratios for the Netherlands, accord-

* See note IV.

† Mean ratio of male and female legitimate births for Europe is.............. 106 to 100, or as 1 to 0·9434 nearly. And for illegitimate... 103 to 100, or as 1 to 0·9708 do.

ing to ten years' observations, collected by the commission of statistics of the kingdom, have the following values :

	Towns.	Country.
1 Birth for	26·07	29·14 Individuals.
1 Death	32·61	43·83.*

It is found, also, that 1 marriage may be calculated on for 132 individuals ; and for each marriage there may be estimated 4 to 5 infants, or more correctly, 4·56 †. This last ratio is the measure of fecundity ; it varies according to the country, as may be seen by the following Table:

Savoy	5·65
Government of Venice..............	5·45
Bohemia	5·27
Moscovy	5·25
Bergamo	5·24
Portugal	5·14
Scotland	5·13
Moravia and Silicia..............	4·81
Netherlands.......................	4·56
France	4·21
Sweden	3·62
England	3·50

The number of births and deaths are not the same in the different months of the year, but they present a *maximum* and a *minimum* ‡.

* See note V.   † See note VI.   ‡ See note VII.

These numbers, notwithstanding the modifications which they ought to undergo from the difference of climates and manners, seem to follow a law which is sufficiently manifested in the observations made during 18 years at Brussels. They have been since verified by more than 13 millions of observations collected by M. Villermé, in different parts of the globe, leaving, therefore, no doubt on the subject.

To form an idea of this law, it will suffice to cast the eyes over the subjoined Table, which is the result of records made at Brussels.

	Deaths.	Births.
January	1172	1040
February	1110	1157
March	1100	1099
April	1068	1079
May	995	989
June	916	956
July	806	901
August	844	903
September	884	940
October	954	949
November	975	968
December	1175	1172

The *minimum* of births which occur in July, seems to obtain earlier, in proportion, as we advance towards the south.

We shall conclude this lesson by a remark relative to a prejudice existing, generally, on the pretended danger of being the 13th at table.

If the probability be required, that out of 13 persons, of different ages, one of them, at least, shall die within a year, it will be found that the chances are about 1 to 1 that one death, at least, will occur. This calculation, by means of a false interpretation, has given rise to the prejudice, no less ridiculous, that the danger will be avoided by inviting a greater number of guests, which can only have the effect of augmenting the probability of the event so much apprehended.

QUESTIONS TO LESSON XV.

What is the object of the Tables of Mortality?

How do you construct a Table of Mortality?

Does this method of construction present any inconvenience?

What is the probability of life?

What is the probability of life, in the principal kingdoms, and in some of the capitals of Europe?

What is mean life?

What is mean life in different places?

How do you calculate the probability of continuing to live a certain number of years, after a given age?

How do you calculate the probability of life at a given age?

At what age is the probability of life at its maximum?

At what age is mean life at its maximum?

What is the probability that two individuals will continue to live to a certain period?

How do you render the law of mortality sensible to the eye?

Are there more males than females born?

What is the ratio of the births, deaths, and marriages, to the population in the Netherlands?

What is the fecundity in the Netherlands?

Are the births and deaths the same in number during the different months of the year?

What do you think of the fear which some persons experience in finding themselves one in a party of 13?

LESSON XVI.

ASSURANCES AND LIFE-ANNUITIES.

ASSURANCE Societies* have for their object, to supply the means, by certain payments, of sheltering men from those chances by which their interests are threatened. Hence assurances on human life, against fire, uncertain seasons, the dangers of the sea, &c.

Assurance on human life, is a sort of contract, by virtue of which a capital or a rent is secured, at the end of a certain time, by means of a sum

* At the first establishment of these societies they were viewed with marked distrust, being considered as mere gambling speculations.

of money paid at once, or a smaller sum paid annually.

The contract receives the name of a policy of assurance; and the sum paid once for all, is termed the price of the assurance, and that paid annually the premium of assurance.

Many modes of assurance exist; we shall exhibit the principal ones.

A person with a view of leaving a capital to his family, after his death, desires to effect an assurance on his life. This assurance may be made, either for a determinate time, as one, two, three years, &c., or for the whole life. In the first case, if the person assured, die before the term stipulated, the inheritors receive the capital; if the person assured, pass the term, they receive nothing; in the second case, the conditions of the policy should be always fulfilled *.

The calculation made, as to the amount to be paid, should be according to mathematical expectation, (see Lesson 2nd,) as equal for one party as for the other, deducting something for the profit of the insurancer. If the price of assuring 100*l.* for one year, be required, supposing the assurer capable of paying the 100*l.*, the probability de-

* In most of the assurance societies, there are, under certain circumstances, cases of exception ; as, for example, when the person assured dies by the hand of justice, or in a duel, or by suicide, &c.

pends on the age of the person assuring. In
equity, then, the sum paid should be equal to the
value of the expectation, multiplied by the probability of obtaining it. If the person assuring be
40 years of age, the probability of death in the
course of the year will be, according to our table,
$\frac{671}{38,670}$, and this fraction, multiplied by 100, gives the
price of the assurance, viz. 1·74 nearly. By the
tables of mortality of France, the result is 1·89.
This is the sum paid to the general assurance
company established at Brussels. The Belgic
and Strangers' Union society, requires 1·87. These
societies follow the Table of Mortality of Dubillard,
to be found in the " Bureau of Longitude," of
France.

It will be seen that the profit of the assurer
is reduced here to the interest of the sum lodged
by the individual assuring. This profit will appear more considerable, if it be remembered that
persons in health only are permitted to assure,
and for whom the probability of dying is, consequently, considerably less than that indicated by
the Tables*.

When a longer term than one year is contemplated, assurance societies calculate the interest of
the money placed in their hands.

* The general assurance company of Brussels, and that
of the union require, in addition to the premium, a fee of
2 fl. 50 c. for preparing the policy of assurance.

The Belgic union society, for example, receives
4·639 fl. as the price of an assurance of 100 fl.
on the whole life, at the age of 40. There is a
certainty of the society being called on to pay
in this case, but at a time more or less remote,
according to the age of the individual assuring.
The calculation reduces itself, then, to the examination of the sum necessary to pay for the value
of 100 fl., with the interest, estimated at 3 or 4
and sometimes at 5 per cent.

Often in place of a capital, an annuity may be
assured to inheritors, having the same value as
the capital, and for which, conseqnently, the same
price is paid, or the same premium.

Two persons, two married persons, for example,
can assure on their united lives, a capital or an
annuity, in favour of the undetermined survivor, or in favour of the one first named. The
case of assurances relates to compound probabilities. The Belgic union society require a premium of 1·60 fl. for an assurance of 100 fl. payable
after the death of two persons whose ages are respectively 40 and 30. The premium is 1·89 fl. if
the assurance of 100 fl. be payable to the younger
of the two persons, if the survivor, and 2·51 if the
survivor be not named.

Assurance societies receive also deposits during
life, donations in favour of infants, weekly savings,

&c.* The payments during life consist in making
one payment, or in giving annual premiums, in
order to acquire a capital or an annuity at a certain
period ; it being necessary to consider, in the calculation, the probability of the life of the person
assuring, and the interest of the money paid.
Gifts in favour of infants, consist in assuring
a capital, or an annuity to an individual, when he
shall have attained a determined age ; this calculation is the same as the former one. The weekly
savings accumulate, with their interests, and are
refunded at the choice of the depositor.

After what has been said, it will be evident that
life assurance societies have for their basis the
laws of human mortality, and the value of interests which produce a certain sum. Hence the
great importance of having the tables of mortality
prepared with care, and the capability even of
establishing the distinction between men and
women, the mortality being generally less amongst
the latter. The tables also require to be revised
as our habits and modes of living become modified.

Assurance societies may be constituted by the
state, by particular societies, or by mutual associations. The two latter are the only ones known

* See Note VIII.

in practice. Particular societies are objects of speculation, where often the advantages of the assurances are immense. The mutual associations, where the assured are the assurers, regulate for themselves, and are interested in procuring full value for all, with the greatest amount of economy. Other societies again exist of a mixed character, societies where the assured are represented, and where they have a part in the profits.

Assurances against the dangers of the sea, against fire, against uncertain seasons, &c., ought to rest on the observation of numerous facts, which are still generally wanting. What has been stated, will assist in calculating the probability of loss or gain, since, strictly speaking, and always subtracting the profits of the assurer, the sum deposited by the person assuring, ought to be worth the product of the value of the property which he assures, by the probability which he has of losing it.

In 1818, there was a bank established at Paris, of savings and of providence, which merits imitation in all countries*. Far from its having been an object of speculation, on the part of its founders, this bank is conducted gratuitously by them; it received from them a donation sufficient to meet

* See l'Annuaire du Bureau des Longitudes de France; for 1838.

the daily expenditure of preparing the accounts, the sole aim being, to offer to the small economists, without expense or risk for the future, an advantageous lodgment, afforded every where else only for sums of some amount.

Assurance societies and savings' banks are establishments eminently useful, when conducted on proper principles; they have, amongst other things, a moral tendency in enabling the provident man to see his economy fructify, not only for his own advantage, but for the benefit of those who are dear to him*. They may be considered as stimulants to labour and good conduct, since the interests of the assured are bound up with them. Assurance societies will be less profitable to governments than the lotteries, but they will confer on them infinitely great honour, and they will become one of the most powerful means for consolidating public tranquillity, while they improve the morality of the people. Governments do not, perhaps, consider sufficiently that in doing good, they interest an infinite number of individuals in their existence†.

QUESTIONS TO LESSON XVI.

What is the object of Assurance Societies?
What is life assurance?

* See Note VIII.   † See Note IX.

What do you understand by a *policy* of assurance?
What do the words *price* and *premium* signify?
What are the principal modes of life assurance?
Give some examples.
Can an assurance be effected on two persons?
What do you understand by life annuities?
What is the principle according to which all assurance societies ought to be regulated?
Are assurance societies of advantage to a state?

LESSON XVII.

ON THE PROBABILITY OF WITNESSES.

ALL that relates to evidence, has been referred to the theory of probabilities: it may be supposed, that out of a given number of depositions, the witness will tell truth a certain number of times.

Example.—If it be observed that out of 10 depositions made by an individual, 9 be uniformly true, the probability of the truth of his deposition is $\frac{9}{10}$, and the probability against it is $\frac{1}{10}$. It is evident that, for a single witness, only two circumstances can be considered, viz. the truth or falsehood of his deposition. If there be two witnesses, four circumstances offer themselves for examination, viz. the truth or falsehood of two depositions, and the two cases where the depositions are contradictory.

Example.—Let the probability that the first witness speaks the truth, be $\frac{9}{10}$; that of a second witness $\frac{7}{8}$; the probabilities against these are $\frac{1}{10}$ and $\frac{1}{8}$. If it be further required to ascertain the compound probabilities for the four different circumstances which may occur, it will be found that, for—

The two witnesses speaking truth $\frac{9}{10} \times \frac{7}{8}$ or $\frac{63}{80}$.

Not speaking truth $\frac{1}{10} \times \frac{1}{8}$ or $\frac{1}{80}$.

The first witness only speaking truth $\frac{9}{10} \times \frac{1}{8}$ or $\frac{9}{80}$.

The second . . . . $\frac{1}{10} \times \frac{7}{8}$ or $\frac{7}{80}$.

Such are the probabilities before the depositions are known, and their sum constitute unity. But when the depositions are made, one of two things occurs, either the depositions agree, or they do not. The estimation of the probabilities refer, then, to relative probabilities, in the following manner:—

1st Case.—Where the witnesses can only both agree by speaking truth, or by lying; now, neglecting all the other chances which it is unnecessary to consider, the preceding calculations shew that there is in this case 63 to 1, where the depositions are not contradictory.

The probability then, that

Both speak truth . . . . . . $\frac{63}{64}$.

Both lie . . . . $\frac{1}{64}$.

2nd Case.—Where the depositions are contra-

dictory, it will only be required to know which of
the two witnesses has told the truth. Now the
chances in favour of the first may be 9, and 7 for
the second, according to what we have seen: thus,
for the probability that

The first speaks truth . . . . . $\frac{9}{16}$.
The second. . . . . $\frac{7}{16}$.

If the probabilities of more than two witnesses
be required, this calculation of compound probabilities may still be depended on; that is to
say, the simple probabilities of all the individual
circumstances, to be considered, ought to be multiplied together.

Example.—It becomes desirable to know the
depositions of four witnesses, whose degrees of
veracity may be represented by the same fraction
$\frac{9}{10}$, and it is asked what probability exists that
their evidences agree, and what that they disagree. It may be observed that the witnesses can
only agree in speaking truth, or in lying: now,
the probability for the first circumstance occurring,
should be $\frac{9}{10} \times \frac{9}{10} \times \frac{9}{10} \times \frac{9}{10}$ or $\frac{6561}{10,000}$, and that for
the second $\frac{1}{10,000}$. Hence, for the probability that
the witnesses agree, there is the fraction $\frac{6562}{10,000}$; and
for the contrary, 1 less the preceding fraction, or
$\frac{3438}{10,000}$.

Suppose, however, that the witnesses have given
their evidence, and that their depositions agree,

there is then no difficulty in knowing whether all spoke truth, or whether all lied. By the preceding calculation it appears, that for the first hypothesis there are 6,561 chances; and for the second 1.

For the probability that

The witnesses speak truth . . . $\frac{6561}{6562}$.

That they do not . . . . . . $\frac{1}{6562}$.

By following out the preceding calculations, we obtain readily the following conclusions:—

1st.—The more the number of witnesses are increased, the greater become the probabilities of their disagreeing.

2nd.—If the depositions have not been contradictory, the more numerous they are, the greater is the probability that they have spoken truth. These conclusions always suppose that the degree of veracity of all the witnesses exceeds $\frac{1}{2}$. If a contrary hypothesis be admitted, the first conclusion still remains true, but the second ought to be modified in the following manner; if the depositions have not been contradictory, the more numerous they are, the greater the probability that they are false.

From what has been said, we ought to be guided in the estimation of the confidence which may be attached to the *traditions* even of ordinary events; for it is obvious, that a statement ought to become as much less probable, as the number of mouths through which it may have passed has increased: and in this case, the last witness only is heard by the individual obtaining the information. When an extraordinary event is under consideration, of which we have been spectators, prudence directs us to look on it as only probable, because the evidence of our senses may deceive us*. The progress of science shews us, in fact, that we may be often exposed to take illusions for realities †.

When this extraordinary event is transmitted to us through a witness, it loses something of its probability, which will become the more feeble, as the event shall have been the subject of a greater number of traditions.

Example.—Four persons, the degree of veracity of each being supposed equal to $\frac{9}{10}$, transmit the recital of an extraordinary event, for the reality of which, had we been ourselves a witness, the chances may have been 1 to 1: what is the probability that the 4th tradition is conformable to truth? It is necessary to multiply $\frac{1}{2}$, the probability of the event, by $\frac{6,561}{10,000}$, (or $\frac{9}{10} \times \frac{9}{10} \times \frac{9}{10} \times \frac{9}{10}$), for the truth of each tradition; we have thus $\frac{6,561}{20,000}$, or $\frac{1}{3}$ nearly. This recalls the expression of the celebrated La Place, relative to the diminution of the probability of traditions, which he compared to the extinction

* See Note X.   † See Note XI.

of the clearness of objects, by the interposition of numerous pieces of glass.

Traditions where the successive evidence is given by simply "*yes* or *no*," must be considered separately; for two false depositions successively made, answer to the truth, the one rectifying the fallacy of the other. The truth will be told here in a different manner, whilst in that which preceded, it could only result from the concurrence of all the witnesses in one expression.

Example.—If it were required to know whether Pierre had been killed in a duel, a third mouth having affirmed the fact? Now, supposing that the fact were true, it may reach us in the four following manners:—

	1.	2.	3.	4.
1st witness	Yes.	yes.	no.	no.
2nd do.	Yes.	no.	yes.	no.
3rd do.	Yes.	yes.	yes.	yes.

The truth may reach us by three truthful assertions; or, out of three modes, by one true assertion and two false ones. In this calculation it is necessary to consider the probabilities of each of these assertions, and in making them equal to $\frac{9}{10}$, $\frac{8}{9}$ and $\frac{7}{8}$, we have the degrees of veracity of the 1st, 2nd, and 3rd witness, hence the following probabilities, in favour of the truth:—

Yes, yes, yes .... $\frac{9}{10} \times \frac{8}{9} \times \frac{7}{8}$ or $\frac{504}{720}$
Yes, no, yes .... $\frac{9}{10} \times \frac{1}{9} \times \frac{1}{8}$ or $\frac{9}{720}$
No, yes, yes .... $\frac{1}{10} \times \frac{1}{9} \times \frac{7}{8}$ or $\frac{7}{720}$
No, no, yes .... $\frac{1}{10} \times \frac{8}{9} \times \frac{1}{8}$ or $\frac{8}{720}$

The sum of these probabilities is .... $\frac{528}{720}$

We may have been led into an error in the following manner:—

	1.	2.	3.	4.
1st witness	Yes.	yes.	no.	no.
2nd do.	Yes.	no.	yes.	no.
3rd do.	No.	no.	no.	no.

And the probabilities of these assertion are:—

Yes, yes, no .... $\frac{9}{10} \times \frac{8}{9} \times \frac{1}{8}$ or $\frac{72}{720}$
Yes, no, no .... $\frac{9}{10} \times \frac{1}{9} \times \frac{7}{8}$ or $\frac{63}{720}$
No, yes, no .... $\frac{1}{10} \times \frac{1}{9} \times \frac{1}{8}$ or $\frac{1}{720}$
No, no, no .... $\frac{1}{10} \times \frac{8}{9} \times \frac{7}{8}$ or $\frac{56}{720}$

The probability that we were led into error by one of these four modes, will be then $\frac{192}{720}$. This fraction and the foregoing together constitute unity.

These examples will suffice to shew how to calculate the probabilities of witnesses and reports, in the more ordinary cases which occur. They teach us, at the same time, how we ought to be on our guard against what we can only know through evidence, or by written reports, which are

K

subject to be altered. When the past time is under consideration, the benefit of printing, which substitutes a single witness for a series of traditions, which might pervert the truth, ought to be a guarantee for the correctness of historical facts; more particularly if those facts have been recounted under the eyes of contemporaries who have passed judgment on them*.

QUESTIONS TO LESSON XVII.

What is the hypothesis by which the theory of probabilities is referred to that of probabilities?

How do you estimate evidence, from one or two witnesses?

---

* We give an example, which will shew how necessary it is, even under such circumstances, to use circumspection. Some days after the battle of Waterloo, a newspaper of the country announced that an august personage had been wounded, taken by the enemy, and subsequently, being rescued from their hands, threw his decorations to his liberators, crying, " My friends, you have merited all !" This fact was repeated, and has been since cited, in many works, as one of the best established historical facts. Our descendants must guard themselves well if they doubt its authenticity, when it has been written, and repeated before our eyes. However, we have seen the author of this story (which he had so innocently imagined) alarmed at the confidence with which it had been received, and at the arguments which may be deduced from it, for the truth of historical facts.

Calculate the probabilities of evidence, before deposition?

Calculate the probabilities of evidence, after deposition?

Give an example of these calculations, when applied to many witnesses.

What are the principles which may be deduced from what has been stated?

How ought traditions to be considered?

What prescribes prudence to us in the case of our being ourselves witnesses to extraordinary events?

How ought we to view extraordinary events, which we only know through the medium of tradition?

How ought we to calculate the probabilites of tradition, or of successive testimonies, simply rendered by yes or no?

Give an example of this calculation.

Is printing useful in establishing the correctness of historical facts?

LESSON XVIII.

ON THE DECISION OF TRIBUNALS AND ELECTIONS.

THE theory of decisions may be referred to that of witnesses, if, out of a similar number of opinions, a judge be always exposed to deceive himself a similar number of times.

Example.—Supposing, that out of 10 opinions there be self-deception once, not an unusual circumstance in such a case : there will be for the probability of a correct judgment $\frac{9}{10}$, and for the contrary $\frac{1}{10}$.

After what has preceded, there will be no difficulty in calculating the probability that one decision will restore unanimity.

For this purpose, it is necessary that all the opinions shall concur in the truth, or be opposed to it.

Example.—What is the probability that three judges pronounce a unanimous decision, if the degree of confidence which each merits be $\frac{9}{10}$? The probability that the three opinions shall be conformable to the truth is compound, and affords the product of $\frac{9}{10} \times \frac{9}{10} \times \frac{9}{10}$ or $\frac{729}{1,000}$; and the probabilities against the truth will be $\frac{1}{10} \times \frac{1}{10} \times \frac{1}{10}$ or $\frac{1}{1,000}$. The probability of unanimity is then $\frac{739}{1,000}$, if the merits of the judgment be not considered.

The degree of confidence which judges obtain, is not a constant quantity, but it varies with the locality, the time, the state of knowledge, political principles, opinions, &c. The preceding calculations can be only admitted, then, with caution. The nature of the case submitted to the judges ought also to exert an influence on their decision: now this decision can be only obtained according to moral proofs, which are but probabilities; for judgment could not be obtained, if mathematical evidence were looked for; but when the proofs are so strong that the product of the error to be feared from their small probability, be inferior to the danger which would result from crime committed with impunity, judgment is commanded for the interests of society.

" This judgment reduces itself, if I mistake not," says La Place, " to the solution of the following question :—Has the proof of the crime of the accused that high degree of probability necessary to induce the citizens to entertain less apprehension of the errors of the tribunal, supposing him to be innocent and condemned, than of new attempts, and these of unfortunate individuals, which would be emboldened by the example of impunity, if he were culpable and acquitted?

" What renders this question almost always insoluble," adds this great geometrician, " is the impossibility of appreciating exactly the probability of the crime, and of fixing that which is necessary for the condemnation of the accused. Each judge, in such a case, is referred to his own discrimination." If we continue to follow La Place, in the examination of the composition of tribunals, we shall find many singular results worthy to be marked.

1st. *The required majority remaining the same, the more the number of judges are increased, the more the probability of error increases.* Thus, the accused would find himself in a less advantageous position before a tribunal of 8 judges, than before one of 6, when a majority of 2 voices suffice

K 3

to condemn him. The probability of the error to be feared, is greater than $\frac{1}{4}$ in the first case; and less than $\frac{1}{4}$ in the second.

In England, unanimity is necessary in a jury of 12 members, for the condemnation of an accused person before the tribunal; in the Chamber of Peers, a majority of 12 voices is required, whatever be the number of judges; now, should there be 212 judges, calculation shews, that the probability of error to be feared will be $\frac{1}{5}$; and only $\frac{1}{8,192}$ in the case of a jury composed of 12 members. It is not, therefore, sufficient that the majority remains the same.

2nd. *When the majority required, is to the minority in a constant geometrical ratio, the more the number of the judges are increased, the more the probability of error diminishes.*

Thus, in the tribunals where two-thirds of the voices are required to condemn, the probability of error to be feared, is about $\frac{1}{4}$ if the number of the judges be 6: it is below $\frac{1}{7}$ if the number be raised to 12.

It is worthy of remark, that with the same laws and the same organization of the tribunals, the same number of acquittals, out of a given number of accused, annually occur. In France, out of 7,234 persons accused in 1825, there were 4,594 condemned; and in 1826, out of 7,613 accused,

4,912 were condemned ; whence it follows, that of 100 accused, 64 were condemned in the former year, and 65 in the latter.

This consequence in the results is deplorable, where one reflects that it may extend to innocent individuals, which it stigmatizes.

The condemnations are somewhat higher in England than in France ; in 1826, out of 100 they were 69 ; in 1825, they were 67 ; and the mean value for 10 years preceding was 67 : the greatest difference being 2 units*. In 1826, the condemnations in Belgium were 84 ; thus, 16 individuals only were acquitted, out of 100 who appeared before the tribunals; a lamentable proportion, but which may, in some measure, be explained by the fact of our (Belgium) not possessing the form of jury trial, which exists amongst our neighbours.

Thus, the same institution of a jury furnishes, in France and in England, nearly the same results notwithstanding the difference in their laws. We (Belgium) have the same laws as France, but what a difference in the application ! Such is, however, the influence of the mode of judging : the number of acquittals in France and England, are double the number of those amongst us.

It is obvious, that the passions will often de-

* See Note XII.

range the calculations relating to witnesses and judgments; it is the same for elections.

Rules may, however, be established, all things being equal, which present the least anomalies : this is the object more especially proposed to be attained in the modes of election.

The most general form of election is by a majority of voices; there appears to be little more to be desired, when two candidates only are the objects ; but if there be many, this method has its inconveniences. Where an elector votes, in fact, for one candidate, he indicates the degree of merit which he attributes to the others.

Borda, proposed to give to the respective merits of the candidates, the values proportional to the rank which would be assigned them by the scrutiny. Suppose, for example, 12 voters and 3 candidates a, b, c; suppose, further, that the scrutiny produces, 7 times, the name of a. in the first rank, and 5 times in the third rank, and that the name of b. appeared only 5 times in the first rank, and 7 times in the second.

If the election is declared, by plurality of voices, a. should have the preference; by the scrutiny of Borda, we shall find that it is b. which has the preference; in fact, there will be found for a.

$$7 \times 3, \text{ plus } 5 \times 1, \text{ or } 26,$$

and for b.

$$5 \times 3, \text{ plus } 7 \times 2, \text{ or } 29,$$

if the scrutiny be made correctly, and if the difference of numbers represent exactly the degree of estimation, this mode of election ought to be preferred; but it would favour combination, which would have the effect of placing a candidate of merit in the last rank, and would open, thus, an advantageous career to mediocrity.

The mode which seems to offer the greatest advantage, consists in comparing the candidates two by two. Thus, in the former example, supposing the votes to be distributed in the following manner:—

$$6 \text{ for } a. \ b. \ c.$$
$$3 \ \dots \ b. \ c. \ a.$$
$$2 \ \dots \ b. \ a. \ c.$$
$$1 \ \dots \ c. \ a. \ b.$$

In comparing the candidates two by two, we find that $a.$ has been preferred to $b.$, 6 plus 1 times; and that $b.$ has been preferred to $a.$ 3 plus 2 times. The candidate $a.$ ought then to succeed before $b.$; in comparing him with $c.$ a similar result obtains.

QUESTIONS TO LESSON XVIII.

Can the theory of judgments be referred to that of evidence?

How do you calculate the probability of rendering a decision unanimous?

Under what circumstances is a judgment demanded for the general interest?

When the required majority remains the same, what ought to be observed relating to the number of the judges?

What ought to be observed, when the majority is to the minority in a constant geometrical proportion?

What are the inconveniences attending the mode of election by a majority of voices?

In what does the mode of election according to Borda consist?

What is the mode of election apparently the most advantageous?

LESSON XIX.

CONCLUSION.

IF we review all that has preceded, we shall be compelled to conclude, that there are but few things of which we can acquire a certainty, and that the greater part of our knowledge, even what relates to essentials, is founded only on probabilities more or less strong. It is then interesting to be enabled to appreciate the value of these probabilities, not only to apply them to particular cases which constantly defeat our calculations, but to elevate us to the knowledge of results which are produced by the same causes always acting, whether they be known, or whether their existence and their mode of action be only revealed by experience. Chance, a mysterious word, which has been too much abused, ought only to be considered as serving to conceal our ignorance; being a phantom which exercises a most absolute empire over the vulgar mind, habituated to consider events only as insulated facts; but which are annihilated before the philosopher, whose eye embraces a long series of events, and whose observation is not deceived by irregularities, which disappear from before his steady view, when he is enabled to take a position sufficiently elevated to seize the laws of nature*. These laws are eternal, immutable as the intelligence from whence they flow; it is not in our power to alter them: but it is the privilege of man to search for, and to seize them, amidst the numberless anomalies which they seem to present. One of the great merits of modern science is, to have been able to subject those grand principles, which appear ready to escape it for ever, to a dependency on numbers; this determination has nothing arbitrary in it; nor does it afford any premium to the subtleties of words, which have been so much abused; by facts, it exists; facts, the value of which it can appreciate.

Thus we have reviewed the calculation of probabilities, which had its origin within two cen-

* See Note XIII.

turies, and which has exhibited its inherent power, in shewing the true theory that ought to regulate games of all kinds; which enters within the dominions of the natural sciences, to indicate the laws of birth, and of mortality; into those of history, to appreciate the value of facts and of traditions;—into the sanctuary of Thémis* to regulate the composition of tribunals, or to apportion the measure of excellence in judgments: it is further seen, under different names, to invade the tribunal and regulate elections; or to enumerate the riches and the wants of the people, by numbers, which no human eloquence can resist. All that can be numerically expressed comes under its jurisdiction; the more science is perfected, the more it tends to re-enter the domain of numbers, which is a sort of centre towards which the sciences seem to converge.

As I have already elsewhere observed, the degree of perfection to which a science has arrived, may be judged of, by the greater or less facility with which it is tried by calculation, conforming so well to the expression of the ancients, which every day's experience confirms: *mundum numeri regunt.*

* See Note XIV.

NOTES.

Note I. Page 44.

A Lottery is a game, consisting of several numbers, of blanks and prizes; these are drawn out of two wheels, one wheel containing the numbers, and the other the blanks or prizes which correspond.

The Romans are said to have invented this game, and Heliogabalus exercised his ingenuity in the arrangement of prizes; one, for instance, being six slaves, while another was six flies.

The first English lottery was drawn in 1569, the proceeds of which were given to the repair of the harbours of the kingdom.

Since the reign of Queen Anne, when lotteries were declared to be public nuisances, they had been licensed by parliament up to 1826; notwithstanding two reports of a committee of the House of Commons, made in 1808, which brought to light most melancholy and disgraceful facts of misery and vice.

Note II. Page 45.

The constitution of modern lotteries is derived from Italy. Geneva became its cradle, and this republic owes the invention to the form of its government. The origin was as follows :—At Geneva, every six months, the election of five senators occurs, to fill the office of magistrates: the names of all those who aspire to the office are put into an urn; and because the competitors amount to the number ninety, hence the origin of that number; and because the office is conferred on the first names which are drawn from the urn, the same spirit of imitation led to limiting each drawing to five fortunate numbers.

The inventor of this system of lottery was Benedetto Gentile, a citizen of Geneva. It was in accordance with this form that the French lottery of 1758 was established, called the Lottery of the Royal Military School: it was suppressed in 1776, and resumed the same year, under the title of the Royal Lottery of France. (*Parisot, Traité du calcul conjectural.*)

Note III. Page 75.

Age.	TABLE OF MORTALITY FOR BELGIUM.				General Table.
	Towns.		Country.		Towns and Country; Males and Females.
	Males.	Females.	Males.	Females.	
Birth.	10,000	10,000	10,000	10,000	100,000
1 Mnth.	8,840	9,129	8,926	9,209	90,396
2 ..	8,550	8,916	8,664	8,988	87,936
3 ..	8,361	8,760	8,470	8,829	86,175
4 ..	8,193	8,641	8,314	8,694	84,720
5 ..	8,069	8,540	8,187	8,587	83,571
6 ..	7,961	8,437	8,076	8,490	82,526
1 Year.	7,426	7,932	7,575	8,001	77,528
18 M.	6,954	7,500	7,173	7,603	73,367
2 Years	6,626	7,179	6,920	7,326	70,536
3	6,194	6,761	6,537	6,931	66,531
4	5,911	6,477	6,326	6,691	64,102
5	5,738	6,295	6,169	6,528	62,448
6	5,621	6,176	6,038	6,395	61,166
7	5,547	6,095	5,939	6,299	60,249
8	5,481	6,026	5,862	6,215	59,487
9	5,424	5,966	5,792	6,147	58,829
10	5,384	5,916	5,734	6,082	58,258
11	5,352	5,873	5,683	6,018	57,749
12	5,323	5,838	5,634	5,960	57,289
13	5,298	5,807	5,589	5,908	56,871
14	5,271	5,771	5,546	5,862	56,467
15	5,241	5,732	5,502	5,796	56,028
16	5,209	5,689	5,456	5,725	55,570
17	5,171	5,645	5,408	5,668	55,087
18	5,131	5,600	5,357	5,608	54,575
19	5,087	5,551	5,302	5,546	54.030
20	5,038	5,500	5,242	5,484	53,450
21	4,978	5,445	5,178	5,421	52,810
22	4,908	5,387	5,109	5,356	52,172
23	4,827	5,326	5,036	5,289	51,465
24	4,740	5,264	4,958	5,222	50,732
25	4,662	5,201	4,881	5,153	49,995
26	4,590	5,138	4,805	5,085	49,298
27	4,523	5,074	4,734	5,016	48,602
28	4,459	5,010	4,673	4,948	47,965

Age.	TABLE OF MORTALITY FOR BELGIUM.				General Table.
	Towns.		Country.		Towns and Country; Males and Females.
Years.	Males.	Females.	Males.	Females.	
29	4,397	4,946	4,620	4,880	47,350
30	4,335	4,881	4,572	4,812	46,758
31	4,275	4,816	4,525	4,744	46,170
32	4,214	4,751	4,478	4,677	45,584
33	4,154	4,686	4,431	4,609	44,996
34	4,094	4,622	4,384	4,542	44,409
35	4,034	4,558	4,337	4,474	43,823
36	3,976	4,490	4,296	4,401	43,236
37	3,918	4,418	4,255	4,329	42,650
38	3,860	4,347	4,215	4,257	42,064
39	3,802	4,277	4,174	4,185	41,476
40	3,744	4,208	4,134	4,112	40,889
41	3,678	4,148	4,090	4,041	40,300
42	3,611	4,088	4,044	3,971	39,697
43	3,544	4,027	3,995	3,901	39,106
44	3,477	3,967	3,943	3,831	38,504
45	3,411	3,907	3,887	3,761	37,900
46	3,352	3,846	3,827	3,701	37,295
47	3,293	3,783	3,767	3,640	36,690
48	3,233	3,720	3,707	3,579	36,084
49	3,174	3,656	3,647	3,519	35,477
50	3,115	3,592	3,588	3,458	34,789
51	3,040	3,520	3,512	3,392	34,153
52	2,962	3,448	3,435	3,323	33,418
53	2,881	3,375	3,358	3,256	32,676
54	2,810	3,300	3,276	3,187	31,930
55	2,739	3,225	3,194	3,118	31,179
56	2,667	3,150	3,111	3,049	30,424
57	2,583	3,080	3,026	2,982	29,656
58	2,499	3,010	2,939	2,912	28,423
59	2,415	2,939	2,851	2,840	27,465
60	2,329	2,862	2,767	2,762	26,875
61	2,239	2,779	2,677	2,677	26,081
62	2,146	2,689	2,587	2,586	25,242
63	2,051	2,595	2,495	2,495	24,356
64	1,956	2,498	2,387	2,405	23,478
65	1,859	2,397	2,277	2.310	22,462
66	1,754	2,292	2,163	2,200	21,362
67	1,649	2,187	2,049	2,086	20,263

Age.	TABLE OF MORTALITY FOR BELGIUM.				General Table.
	Towns.		Country.		Towns and Country; Males and Females.
Years.	Males.	Females.	Males.	Females.	
68	1,556	2,085	1,942	1,983	19,219
69	1,466	1,983	1,835	1,875	18,175
70	1,372	1,864	1,713	1,758	17,017
71	1,279	1,741	1,587	1,642	15,860
72	1,184	1,627	1,474	1,530	14,749
73	1,087	1,514	1,358	1,420	13,638
74	989	1,389	1,236	1,300	12,461
75	891	1,261	1,114	1,182	11,273
76	806	1,134	996	1,061	10,120
77	721	1,011	882	940	9,014
78	631	900	770	832	7,910
79	541	789	664	723	6,853
80	463	682	566	619	5,867
81	394	585	482	535	5,031
82	332	495	414	460	4,299
83	273	411	353	390	3,627
84	225	346	294	323	3,016
85	184	289	239	262	2,464
86	150	239	199	211	1,989
87	120	192	152	168	1,585
88	93	150	117	132	1,233
89	69	116	88	97	924
90	49	86	67	71	682
91	37	65	48	54	510
92	28	47	38	40	387
93	18	33	27	32	282
94	11	24	20	24	207
95	9	18	14	18	153
96	5	12	10	12	105
97	4	8	7	7	67
98	2	4	4	4	39
99	1	2	2	2	20
100	....	1	1	1	10
101	....	....	....	....	5
102	....	....	....	....	2
103	....	....	....	....	1
104	....	....	....	....	......

M. Quetelet, " Sur l'Homme," &c., Vol. I. p. 170.

Note IV. Page 80.

" The limiting age of manhood," says Mr. Edmonds, in his work on Mortality, " is variable for different classes of the population. In England, I would place it for a city population, at 55; for the general population, at 52; and for the monied population, at 49 years of age.

" The worst kind of life, or the severest mortality, is to be looked for in the poorest class of a city population, and in the highest class of the monied, or non-labouring portion of the community.

" The city table represents the greatest rate of mortality ever shewn to exist in any class of monied life, and it is found perfectly applicable to the English Peerage."

Note V. Page 81.

M. Quetelet, in his recent work, "Sur l'Homme et le développement de ses Facultés," considers it to be ascertained,—1st, " That the population increases according to a geometrical progression, if no obstacles be presented to its development: and 2dly, That the impediment, or the sum of the obstacles to the development, is, all things being otherwise equal, as the squares of the velocity with which the population tends to increase."

The result of statistical inquiries of this nature exhibits strongly the influence of civilization, not only in adding to the happiness of man, but in prolonging life.

" It appears to be well established," says M. Quetelet, " that in the country where civilization has made the greatest progress, there the mortality is least.

" England, for example, has placed itself in a position so advantageous, as to have fixed the attention of the learned, who have occupied their minds on the theory of population.

" If we examine into what has been the state of mortality since the commencement of the 18th century, we shall find, according to two of the best authorities, Marshall and Rickman, the following result:—

Years.	Inhabitants for one death.
1700	43
1750	42
1776 to 1800	48
1806 to 1810	49
1816 to 1820	55
1826 to 1830	51 "*

" The changes which have occurred in large towns, merit attention. In 1697, for example, the total number of deaths in London, was 21,000; however, in the following century, 1797, the number was only 17,000, notwithstanding the increase of population.

" These advantages have been more remarkable within 50 or 60 years, though the town has extended its limits, and increased its population with so much rapidity. In the middle of the last century, the annual mortality was still 1 in 20. In 1821, it was not more than 1 in 40 : thus diminishing the mortality one-half. The towns of Manchester, Liverpool, and Birmingham, present nearly the same decrease as that of London.

" France, like England, has experienced a diminution of mortality, if reference be made to ancient documents. According to M. Villermé, there was, in 1791, 1 death out of

* M. Quetelet, " Sur l'Homme," &c., Vol. I. p. 251.

29 inhabitants; in 1802, 1 out of 30; and at present, 1 out of 40."

M. Moreau de Jonnès, in a notice on the mortality of Europe, has given the following Table, which tends equally to prove the influence of civilization, on the number of deaths.

Countries.	Years.	One Death out of	Years.	One Death out of
Sweden	1754 to 1768	34	1821 to 1825	45
Denmark	1751 — 1754	32	1819	45
Germany	1788	32	1825	45
Prussia	1717	30	1821 to 1824	39
Wurtenburg	1749 to 1754	31	1825	45
Empire of Austria..	1822	40	1825 to 1830	43
Holland	1800	26	1824	40
England..........	1690	33	1821	58
Great Britain......	1785 to 1789	43	1800 to 1804	47
France..........	1776	25·5	1825 — 1827	39·5
Canton de Vaud ...	1756 to 1766	35	1824	47
Lombardy	1767 — 1774	27·5	1827 to 1828	31
Roman States	1767	21·5	1829	28
Scotland..........	1801	44	1821	50*

M. Moreau de Jonnès has also given a table of mortality for different latitudes; shewing the influence of locality, and high temperature.

Under the Latitude of	Places.	One Death per
6° 10′	Batavia	26 Inhabitants.
10° 10′	Trinidad	27 ......
13° 54′	St. Lucia	27 ......
14° 44′	Martinique	28 ......
15° 59′	Guadeloupe	27 ......
18° 36′	Bombay..........	20 ......
23° 11′	Havannah........	33 ......

* M. Quetelet, " Sur l'Homme," Vol. I. p. 255.

We are, however, furnished with more accurate information, connected with the mortality of the British West India Islands, through the valuable Report made to Parliament, "on the Sickness, Mortality, and Invaliding among the Troops in the West Indies," by Captain Alexander M. Tulloch and Mr. Henry Marshall—May, 1838. A report replete with deep interest to humanity, and of inestimable value as an example, of statistical accuracy and research. From that report the following Table has been formed.

TABLE OF THE MORTALITY IN THE WEST INDIES.

Longitude.	Latitude.	Place.	Inhabitants for one death of the white soldiers.	Inhabitants for one death of the black.	Fall of Rain in Inches.
56° to 60° West	6° 10' to 8° 0' North	British Guiana	12·00	24·63	157
60° 30' to 61° 20' West	9° 30' to 10° 51' North	Trinidad	9·43*	25·00	—
60° to 30' West	11° 16' North	Tobago	6·5	29·4	—
61° 20' to 61° 35' West	11° 58' to 12° 20' North	Grenada	16·00	35·7	65
60° 37' West	13° 10' North	St. Vincent's	18·00	27·7	70 to 80
59° 41' West	13° 5' North	Barbadoes	17·00	21·73	—
60° 58' West	13° 50' North	St. Lucia	8·4	24·00	Very moist.
61° 15' West	15° 25' North	Dominica	7·2	25·00	—
62° 7' West	17° 3' North	Antigua	25	34·5	45
62° 13' West	16° 47' North	Montserrat	25	34·5	—
62° 40' West	17° 18' North	St. Christopher	About 13·4	22	—
63° 3' West	17° 14' North	Nevis			
64° 39' West	18° 20' North	Tortola			
76° 0' to 78° 40' West	17° 35' to 18° 8' North	Jamaica	7·8	33·3	—

The mortality of the whites may be taken at 93½ per 1000 of the strength, or about an eleventh part of the force, annually; while the mortality at home is about 15 per 1000, or 1 in 66·66.

   * Deducting the years 1817, 18, 19, when the numbers were 16·666.

Many opinions hitherto entertained with regard to the nature and influence of the essential causes of sickness, and mortality from continued exposure to a high temperature, appear from the detail afforded in this valuable report to be altogether refuted, and shew them to have been adopted on very inadequate evidence.

" The range of the thermometer, in Antigua and Barbadoes," say the authors of the report, " is rather higher than in Dominica, Tobago, Jamaica, or the Bahamas ; yet we find that the troops in the latter stations suffer nearly three times as much as those in the former."

If elevated temperature were an essential cause of mortality, we might expect it in every year to produce similar effects ; whereas, the contrary is the case, nor can excess of moisture be considered as in itself adequate, for if it were so, " we might expect it to attain its maximum in those situations where the fall of rain is greatest; whereas the average mortality of the troops in Jamaica is at least double that which prevails among those in British Guiana, though the quantity of rain which falls in that island is little more than half as great.

" It must be further remembered, that the excess of moisture is not confined to the West Indies, but is a general characteristic of all tropical regions; and were it so productive of disease in the western hemisphere, the same effects might be expected to ensue from it in the east; whereas, on the contrary, the Malabar coast, which is deluged by rain for six months in the year, is generally one of the most healthy quarters in the Madras Presidency.

" Though heat and moisture be not the primary causes of fever, however, it is highly probable their operation tends, in some measure, to increase its intensity. The Tables shew that the greatest number of admissions into hospital, and deaths, has, on an average of a series of years, (though not

uniformly, or equally in each year,) taken place in those
months when the greatest degree of heat was combined with
the greatest moisture; and it may be observed, as a striking
exemplification of this fact, that as the sun proceeds north-
ward in the ecliptic, carrying heat and moisture in his train,
the period generally termed the unhealthy season is later in
the northern colonies than in those to the south." This
effect is also observed in the east, and in a large portion of
the northern temperate zone. " In the Mediterranean stations particularly, the admissions into hospital, and deaths,
among the troops, average nearly twice as high between
July and October, as during any other months of the year.
Even in Canada the same peculiarity is observable, though
not in so marked a degree; and conversely in stations south-
ward of the equator, that period of the year, which on the
north of the line is the most unhealthy, becomes, in the
south, the most salubrious, in consequence of the seasons
being reversed."

The electrical condition of the atmosphere, in the West
Indies, is suggested as the cause of epidemic.

" The instances of Fort St. George, in Tobago, Morne
Fortune, at St. Lucia, and Morne Bruce, at Dominica, demonstrate that mere elevation to the height of 600 or 700
feet, instead of procuring a healthy position, seems rather to
have the reverse tendency.

" 2,000 to 2,500 feet elevation, either wholly exempts from
remittent fever, or modifies its form, so that the mortality at
that elevation, from all causes, does not, on the average of
a series of years, materially exceed that to which an equal
number of European troops would be subject in the capital
of their native country. The diseases of the tropics seem,
like the vegetable productions of the same regions, to be
restricted to certain altitudes, and particular degrees of temperature." " The researches of Humboldt on this subject

have tended to establish, that yellow fever is never known beyond the height of 2,500 feet, so that the nearer this boundary can be approached the more likely is the health of the troops to be secured."

" Where sufficient elevations cannot be obtained, the localities which appear least subject to yellow fever, are low sandy tongues of land, or peninsulas jutting into the sea, and not shut in by any high ground immediately in the rear."

The influence of climate would, however, be more justly considered in its effects on the aborigines. " So far as statistical inquiries have extended," say the authors of the report referred to, " there is no country, either temperate or tropical, in which the mortality among the indigenous civil inhabitants, between the ages of 20 and 40, seems materially to exceed 15 per 1,000 annually (1 in 66·66); and probably there is no country where troops composed of the indigenous inhabitants are subject to a higher rate. As an instance, the 'Malta Fencibles, composed of the natives of that island, the average mortality does not exceed 9 per 1,000 annually (1 in 111·1); and among the Cape corps, composed of the aboriginal inhabitants of southern Africa, it does not exceed 11 per 1,000 (1 in 90, nearly): while among the Madras native troops it does not exceed 13, (1 in 76·9)—nor among those of Bengal is it more than 11 per 1,000 (1 in 91, nearly)."

The inference, then, as regards the health of the negro troops is, that the West Indies proves also prejudicial to them, in as far as the mortality exceeds that of native troops of other countries. On the average of the last 20 years, 40 have died out of every 1,000, (1 in 25,) annually. Nor does a high mortality amongst negro troops appear to be limited to the West Indies. In the Mauritius the mortality is nearly the same. In Ceylon, where they were employed

M

in the colonial corps, they became nearly extinct in a few years. In Gibraltar, where the fourth West India regiment was stationed for two or three years, 62 per 1,000 died annually.

At Sierra Leone, on the sea coast of their own continent, the mortality has averaged not less than 28 per 1000 (1 in 35·7).

" This demonstrates beyond a doubt, that the constitution of the negro can be but little fitted to adapt itself to foreign climates."

The proportion in the mortality amongst the slave population throughout the West Indies, exclusive of Jamaica, is 1 in 33 of the population annually, whereas in most other countries, the mortality amongst the indigenous inhabitants, is only 1 in 45 to 1 in 50 annually; and this extra mortality among the negro population falls almost entirely upon the adults, negro children being in general remarkably exempt from disease.

To the foregoing observations may be added the following highly interesting Tables.

| | Admissions into Hospital. | Deaths. |
	Annual ratio per 1000 of mean strength.	Annual ratio per 1000 of mean strength.
Fevers	717	36·9
Eruptive fevers	0 $\frac{2}{10}$	....
Diseases of the lungs	115	10·4
„ of the liver	22	1·8
„ of the stomach and bowels	421	20·7
„ of the brain	28	3·7
Dropsies	7 $\frac{8}{10}$	2·1
Rheumatic affections	49	
Venereal	35	
Abscesses and ulcers	204	
Wounds and injuries	129	2·9
Punished....................	50	
Diseases of the eyes	89	
„ of the skin	6	
All other diseases	30	
Total	1,903	78·5

WINDWARD AND LEEWARD COMMAND.

Age.	Aggregate strength at each age in returns of seven years.	Total deaths at each age in returns of seven years.	Annual ratio of deaths per thousand living at each age.
Under.. 18	244	1	4
18 to 25	12,372	613	50
25 33	13,633	1,004	74
33 40	3,555	346	97
40 50	609	75	123
Total	30,413	2,039	67

JAMAICA COMMAND.

Age.	Aggregate strength at each age in returns of seven years.	Total deaths at each age in returns of seven years.	Annual ratio of deaths per thousand living at each age.
Under.. 18	88	5	57
18 to 25	8,059	562	70
25 33	6,607	705	107
33 40	1,547	203	131
40 50	352	45	128
Total	16,653	1,520	91

" The numbers under 18 years of age are too few to found thereon any definite conclusions ; but they seem to warrant the inference, that in the West Indies, as in this country, such persons are less subject to mortality than those at a more advanced period of life. The influence of age, in increasing the mortality among the other classes, will best be comprehended by a comparison of the above results with what occurs among the civil population of Great Britain, at the same ages :—

	Annual ratio of mortality per thousand living at the following ages.			
	18 to 25	25 to 33	33 to 40	40 to 50
In civil life in England, by Carlisle Tables..........	7	8·9	10·7	14·1
Among troops in the Windward and Leeward command.................	50	74·	97·	123·
Among troops in Jamaica command	70	107·	131·	128·

" Thus, instead of the mortality among our troops in the West Indies decreasing with the advance of age, as has been the general impression, it increases with infinitely greater rapidity than in this country; and the same has been found to take place at every station, whether temperate or tropical, to which similar investigations have extended.

" Persons who make themselves acquainted with the laws which regulate mortality, will see little reason for adhering to the general belief that soldiers, either in the West Indies or other tropical climates, are more liable to it at an early than at an advanced period of life. We are quite aware that these conclusions are by no means in unison with those generally entertained in the army ; but as they rest on facts

the accuracy of which cannot be disputed, we trust they
will be sufficient to displace hypothetical opinions, which
have principally originated in the want of accurate statistical information on this subject.

" It may perhaps tend to facilitate conviction on this head,
that it has recently been ascertained, that the mortality from
fever, in this country, is materially influenced by the age of
the patient, as will be seen from the following statement of
the number who died out of every hundred cases treated,
at the following ages, in three of the principal London hospitals :—

Age of Patients.	Died per cent.	Died per cent.	Died per cent.
20 to 30	6·8	19·6	14·3
30 .. 40	8·8	26·4	19·5
40 .. 50	11·7	34·8	26·2

See Medical Almanac for 1837.

" Now as fever occasions half of the deaths in the Windward
and Leeward command, and more than three-fourths of those
in Jamaica, it is impossible, if its fatal tendency is regulated
by the same laws in regard to age as in this country, that
the mortality there, at an early period of life, can be the
same as at a more advanced age."

M. Quetelet gives the following Table in the work alluded to, Vol. I. p. 156.

Towns.	Inhabitants for one death.	Inhabitants for one birth.
London	46·0 ⎱ 46·4	40·8 ⎱ 35·2
Glasgow	46·8 ⎰	29·5 ⎰
Madrid	36·0	26·0
Livourne	35·0 ⎫	25·5 ⎫
Moscow	33·0	28·5
Lyons	32·2	27·5
Palermo	32·0 ⎬ 32·3	24·5 ⎬ 27·0
Paris	31·4	27·0
Lisbon	31·1	28·3
Copenhagen	30·3	30·0
Hamburgh	30·0 ⎭	25·5 ⎭
Barcelona	29·5 ⎫	27·0 ⎫
Berlin	29·0	21·0
Bordeaux	29·0	24·0
Naples	28·6	23·8
Dresden	27·7	23·0
Amsterdam	27·5 ⎬ 26·6	26·0 ⎬ 24·2
Brussels	25·8	21·0
Stockholm	24·6	27·0
Prague	24·5	23·3
Rome	24·4	30·6
Vienna	22·5 ⎭	20·0 ⎭
Venice	19·4 ⎱ 18·7	26·5 ⎱ 23·2
Bergamo	18·0 ⎰	20·0 ⎰

Note VI. Page 81.

M. Benoiston de Chateauneuf, quoted by M. Quetelet, in his notice on the fecundity in Europe at the commencement of the 19th century, (Annales des Sciences Naturelles,) has the following interesting observations :—

" If Europe be divided into two climates, one of which, supposed to commence in Portugal and to terminate in the Netherlands, and extending, therefore, from the 40th to the 50th degree, will represent the South ; while the other, extending from Brussels to Stockholm, or from the 59th to the 67th degrees, will represent the North ; it is found that in the first, 100 marriages give 437 births, (or 4·37 births for one marriage,) and that, in the second, the same number of unions produce only 430, or 4·3 births for one marriage.

" The difference becomes still greater, if the two extreme temperatures only be compared. In Portugal each marriage produces 5·10 infants. In Sweden 3·62 only. Lastly, without leaving France, this observation finds corroboration."

" Fecundity," says Moheaux, " increases from North to South. In the latter, the mean number of births to a marriage is 5·03 ; and in the northern provinces, it is only 4·64."

The same proportion exists now as did fifty years ago.

In order, however, to obtain an idea of the value of this fecundity, it must be compared with the population, and with the number of deaths, for it has been observed, that the greater the fecundity, the greater also the mortality. " England, and the province of Guanaxuato, in Mexico, offer remarkable examples, thus :—

INHABITANTS.			
STATES.	For a Marriage.	For a Birth.	For a Death.
England	134·00	25·00	58·00
Republic of Guanaxuato	69·76	16·08	19·70

Quetelet, " Sur l'Homme," Vol. I. p. 151.

" An enlightened people," continues M. Quetelet, " will be less disposed to contract alliances, and to form for themselves a future charged with inquietudes, in a country where each individual obtains, with difficulty, sufficient for his own sustentation." " When man no longer reasons, as is the case where he is demoralized by misery, and where he lives from hand to mouth, the cares of a family touch him less than his own selfishness, and urged by the sole gratification of the moment, he reproduces without any anxiety for the future, resigning to that Providence which had sustained him, the care of offspring to which he gives existence." To use the indefinite language of the Irish people, " Sure God is good ! "

" Habits of order and foresight ought to exercise a great influence on the number of marriages, and consequently on that of the births. The man whose condition is unsettled, if he be guided by reason, fears to bring upon his family the vicissitudes of fortune to which he is himself exposed : hence, many economists with reason argue, that the most efficacious means of preventing a superabundance of population in a country, is to spread the lights of knowledge, and the sentiments of order and foresight." " By restraining fecundity," says Mr. Edmonds, " there is no class of men, however poor, who may not become rich, and command all the real enjoyments of life."

M. Quetelet elsewhere remarks, in the work before quoted,
" that a country becomes most prosperous, when it *gives life
to few* citizens, but *preserves many*.   The increase is altogether to its advantage; for if the fecundity be less, the useful men are most numerous, and generations are not renewed
with so much rapidity as to prove detrimental to the nation.
Man, during his early years, lives at the cost of the society;
he contracts a debt which he ought to liquidate some day
or other; and if he perish before he has done so, his existence will have been rather a burden than a benefit.

" It was found, that from birth to the age of 12 to 16
years, the whole charge of supporting an infant in the hospitals of the kingdom of the Netherlands amounted, in 1821,
to about 1,110—say 1,000 francs, (41*l*. 13*s*. 4*d*.,) and this
sum is not too much for France.   Every individual that
escapes from infancy, has contracted a debt the minimum of
which is 1,000 francs, a sum paid by society for the sustentation of the infant abandoned to its charity.   Now there is
annually born in France above 960,000 infants, of which
nine-twentieths are carried off before they have the power
of rendering themselves useful; these nine-twentieths or
432,000 unfortunates, may be looked upon as so many
stranger friends, who, without fortune, without industry,
come to take part in the consumption, and who subsequently retire without leaving any other traces of their
passage but sad adieus and eternal regrets.   The expenditure
which they occasion, without reckoning the time bestow-
ed on them, is represented by the enormous sum of 432
millions of francs (about 18,000,000*l*.).   If, on the other
hand, the griefs be considered, excited by such deplorable
losses, griefs which no human sacrifice can compensate, it
will be felt how far this subject is worthy to occupy the
mind of the statesman and philosopher, the true friends of

mankind. It cannot be too often repeated, that the pros-
perity of states ought to consist less in the multiplication
than in the conservation of the individuals of which they
are composed."*

"The production of illegitimate children merits, for many
reasons, a particular attention; on political grounds above
all, it should be made an object of the most serious inquiry,
tending as it does to spread throughout society an increasing
number of individuals, who are deprived of the means of
existence, and who necessarily become a charge on the state.
With an organization generally feeble, these individuals rarely
attain to maturity, as we shall shew, so that they do not
offer the same hope of one day affording compensation for
the sacrifices made for them."

According to Mr. Babbage†—

For 1000 legitimate infants.		For one illegitimate infant.
In France	69·7 illegitimate.	14·3 legitimate.
Kingdom of Naples ..	48·4	20·6 ......
Prussia	76·4	13·1 ......
Westphalia	88·1	11·4 ......
Towns of Westphalia.	217·4	4·6 ......
Montpelier	91·6	10·9 ......

"For about fifty years, at Stockholm, Gottingen, and at
Leipsic, the sixteenth part of the births have been illegiti-
mate; the fourth part at Cassel; and the seventh at Jena."
— *Casper Beitrage, &c.*

---

* Quetelet, "Sur l'Homme," Vol. I. p. 152.

† Letter to the Right Hon. T. P. Courtenay.

At Berlin the following results have been obtained:—

From 1789 to 1793	26,573 births, of which 2,824 illeg^{te}. =9 to 1
1794 .. 1798	30,165 .... 3,006 .... =9..1
1799 .. 1803	31,538 .... 3,800 .... =8..1
1804 .. 1808	30,459 .... 4,941 .... =6..1
1819 .. 1822	26,971 .... 4,319 .... =6..1
From 1789 to 1822	145,705 ....18,890 .... =7 to 1

The illegitimate births have therefore increased within late years. The numbers, as determined for Paris, according to the " Annuaires du Bureau des Longitudes," are—

Years.	Legitimate.	Illegitimate.	One illegitimate for legitimate births.	Ratio to population.
1823	27,070	9,806	2·76	$\frac{1}{91}$
1824	28,812	10,221	2·82	$\frac{1}{87}$
1825	29,253	10,039	2·91	$\frac{1}{89}$
1826	29,970	10,502	2·85	$\frac{1}{85}$
1827	29,806	10,392	2·86	$\frac{1}{86}$
1828	29,601	10,475	2·81	$\frac{1}{83}$
1829	28,721	9,953	2·88	$\frac{1}{89}$
1830	28,587	10,007	2·85	$\frac{1}{89}$
1831	29,530	10,378	2·83	$\frac{1}{86}$
1832*	26,283	9,237	2·84	$\frac{1}{96}$
Mean..	287,633	101,010	2·84	$\frac{1}{96}$

" Thus, for twenty-eight births there are almost exactly ten illegitimate; this ratio is, I think, the most unfavourable known."†

---

* In these numbers there has not been included 1,099 and 1,065 infants, recognized and legitimatized after their birth.

† Quetelet, Vol. I. p. 118.

The fifth column has been added, that a comparison may be made with the following Table, the population of Paris being supposed 890,431, according to Malte Brun*.

---

* Geography, Vol. VIII. p. 149.

No. 4.—An Account of the ANNUAL AVERAGE NUMBER OF
25th March, 1836, with their proportion to the Population of
to the Parish, in the years ended 25th March 1835 and 1836
together with the number of Bastards affiliated and the decrease
and Wales.

COUNTIES.	Population in 1831.	Average annual number of Bastards chargeable to the parish in the two years ended 25th March 1836.	Their proportion to the Population in 1831.	Number of Bastards chargeable to the parish in the year ended 25th March 1835.	Number of Bastards chargeable to the parish in the year ended 25th March 1836.
Bedford	95,483	326	$\frac{1}{293}$	377	274
Berks	145,389	786	$\frac{1}{188}$	925	646
Bucks	146,529	670	$\frac{1}{219}$	748	591
Cambridge	143,955	637	$\frac{1}{226}$	644	630
Chester	334,391	2,043	$\frac{1}{164}$	2,159	1,927
Cornwall	300,938	1,028	$\frac{1}{293}$	1,059	997
Cumberland	169,681	1,433	$\frac{1}{118}$	1,532	1,333
Derby	237,170	1,157	$\frac{1}{205}$	1,225	1,088
Devon	494,478	2,430	$\frac{1}{203}$	2,537	2,322
Dorset	159,252	886	$\frac{1}{180}$	973	799
Durham	253,910	1,036	$\frac{1}{245}$	1,113	959
Essex	317,507	1,184	$\frac{1}{268}$	1,217	1,150
Gloucester	387,019	1,756	$\frac{1}{220}$	1,977	1,534
Hereford	111,211	1,026	$\frac{1}{108}$	1,085	966
Hertford	143,341	360	$\frac{1}{398}$	418	302
Huntingdon	53,192	235	$\frac{1}{226}$	246	224
Kent	479,155	2,703	$\frac{1}{177}$	2,966	2,440
Lancaster	1,336,854	3,143	$\frac{1}{425}$	3,345	2,940
Leicester	197,003	712	$\frac{1}{277}$	738	686
Lincoln	317,465	1,973	$\frac{1}{161}$	2,041	1,905
Middlesex	1,358,330	4,526	$\frac{1}{300}$	5,192	3,860
Monmouth	98,130	371	$\frac{1}{264}$	387	355
Norfolk	390,054	1,877	$\frac{1}{208}$	1,938	1,815
Northampton	179,336	827	$\frac{1}{217}$	869	785
Northumberland	222,912	707	$\frac{1}{315}$	759	654
Nottingham	225,327	1,043	$\frac{1}{216}$	1,132	953
Oxford	152,156	917	$\frac{1}{163}$	1,001	833
Rutland	19,385	80	$\frac{1}{242}$	84	75

BASTARDS chargeable to the Parish, for the two years ended 1831; also, an account of the number of Bastards chargeable respectively, with the decrease in 1836 as compared with 1835; in affiliation during the same period in each County in England

Decrease.	Decrease per cent.	Number of Bastards affiliated in the year ended 25th March 1835.	Number of Bastards affiliated in the year ended 25th March 1836.	Decrease.	Decrease per cent.
103	27	83	21	62	75
279	30	89	50	39	44
157	21	96	34	62	65
14	2	230	190	40	17
232	11	329	202	127	39
62	6	365	263	102	28
199	13	113	66	47	42
137	11	204	91	113	55
215	8	453	293	160	35
174	18	311	218	93	30
154	14	163	130	33	20
67	6	198	78	120	61
443	22	337	183	154	46
119	11	235	195	40	17
116	28	45	33	12	27
22	9	66	60	6	9
526	18	216	114	102	47
405	12	1,206	372	834	69
52	7	136	87	49	36
136	7	451	359	92	20
1,332	26	318	110	208	65
32	8	60	50	10	17
123	6	537	426	111	21
84	10	325	238	87	27
105	14	164	111	53	32
179	16	168	85	83	49
168	17	171	104	67	39
9	11	26	15	11	42

COUNTIES.	Population in 1831.	Average annual number of Bastards chargeable to the parish in the two years ended 25th March, 1836.	Their proportion to the Population in 1831.	Number of Bastards chargeable to the parish in the year ended 25th March, 1835.	Number of Bastards chargeable to the parish in the year ended 25th March, 1836.
Salop	222,938	2,037	$\frac{1}{109}$	2,154	1,919
Somerset	404,200	2,260	$\frac{1}{179}$	2,408	2,112
Southampton or Hants	314,280	1,317	$\frac{1}{238}$	1,484	1,150
Stafford	410,512	2,255	$\frac{1}{162}$	2,500	2,009
Suffolk	296,317	1,558	$\frac{1}{190}$	1,652	1,463
Surrey	486,334	2,082	$\frac{1}{234}$	2,233	1,931
Sussex	272,340	1,481	$\frac{1}{184}$	1,691	1,270
Warwick	336,610	1,551	$\frac{1}{217}$	1,664	1,437
Westmoreland	55,041	619	$\frac{1}{89}$	631	606
Wilts	240,156	1,753	$\frac{1}{137}$	1,871	1,635
Worcester	211,365	1,058	$\frac{1}{200}$	1,154	961
York, East Riding	204,253	1,567	$\frac{1}{130}$	1,647	1,486
North Riding	190,756	1,506	$\frac{1}{127}$	1,558	1,453
West Riding	976,350	3,879	$\frac{1}{252}$	4,141	3,617
Totals of England	13,091,005	60,795	$\frac{1}{215}$	65,475	561,092
WALES.					
Anglesey	48,325	238	$\frac{1}{203}$	236	240
Brecon	47,763	235	$\frac{1}{203}$	235	235
Cardigan	64.780	400	$\frac{1}{162}$	415	384
Carmarthen	100,740	837	$\frac{1}{120}$	845	828
Carnarvon	66,448	296	$\frac{1}{224}$	292	299
Denbigh	83,629	609	$\frac{1}{137}$	643	574
Flint	60,012	355	$\frac{1}{169}$	374	336
Glamorgan	126,612	322	$\frac{1}{393}$	312	331
Merioneth	35,315	217	$\frac{1}{163}$	217	216
Montgomery	66,482	994	$\frac{1}{67}$	1,007	980
Pembroke	81,425	863	$\frac{1}{94}$	831	894
Radnor	24,651	417	$\frac{1}{59}$	416	417
Totals of Wales	806,182	5,783	$\frac{1}{139}$	5,823	5,734
Totals of England and Wales	13,897,187	66,578	$\frac{1}{209}$	71,298	61,826

Increase.	Decrease.	Increase per cent.	Decrease per cent.	Number of Bastards affiliated in the year ended 25th March, 1835.	Number of Bastards affiliated in the year ended 25th March, 1836.	Decrease.	Decrease per cent.
..	235	..	11	345	213	132	28
..	296	..	12	575	409	166	29
..	334	..	23	135	73	62	46
..	491	..	20	274	157	117	43
..	189	..	11	311	209	102	33
..	302	..	14	263	158	105	40
..	421	..	25	164	34	130	79
..	227	..	14	305	162	143	47
..	25	..	4	69	55	14	20
..	236	..	13	356	277	79	22
..	193	..	7	127	62	65	51
..	161	..	10	233	200	33	14
..	105	..	7	235	184	51	22
..	524	..	13	757	405	352	46
..	9,383	..	14	11,244	6,776	4,468	40

WALES.

Increase.	Decrease.	Increase per cent.	Decrease per cent.	1835	1836	Decrease.	Decrease per cent.
4	..	2	..	21	14	7	33
..	..	..	..	43	43	..	..
..	31	..	7	96	96	..	..
..	17	..	2	123	98	25	20
7	..	2	..	133	97	36	27
..	69	..	11	158	119	39	25
..	38	..	10	85	55	30	35
19	..	6	..	54	33	21	39
..	1	..	..	62	51	11	18
..	27	..	3	106	73	33	31
63	..	8	..	146	139	7	5
1	..	..	..	110	92	18	16
94	183	2	3	1,137	910	227	20
94	9,566	..	13	12,381	7,686	4,695	38

N 3

If we consider the several Counties according to the ratio of morality, they will present themselves in the following order.

ENGLAND.

Number.	Counties.	Ratio	Number.	Counties.	Ratio
1	Lancaster	$\frac{1}{425}$	19	Norfolk	$\frac{1}{208}$
2	Hertford	$\frac{1}{308}$	20	Derby	$\frac{1}{205}$
3	Northumberland	$\frac{1}{313}$	21	Devon	$\frac{1}{203}$
4	Middlesex	$\frac{1}{300}$	22	Worcester	$\frac{1}{200}$
5	Bedford } Cornwall	$\frac{1}{293}$	23	Suffolk	$\frac{1}{190}$
			24	Berks	$\frac{1}{188}$
6	Leicester	$\frac{1}{277}$	25	Sussex	$\frac{1}{184}$
7	Essex	$\frac{1}{268}$	26	Stafford	$\frac{1}{182}$
8	Monmouth	$\frac{1}{264}$	27	Dorset	$\frac{1}{180}$
9	West Riding, York	$\frac{1}{252}$	28	Somerset	$\frac{1}{179}$
10	Durham	$\frac{1}{243}$	29	Kent	$\frac{1}{177}$
11	Rutland	$\frac{1}{242}$	30	Oxford	$\frac{1}{166}$
12	Hants	$\frac{1}{238}$	31	Chester	$\frac{1}{161}$
13	Surrey	$\frac{1}{234}$	32	Lincoln	$\frac{1}{161}$
14	Cambridge } Huntingdon	$\frac{1}{226}$	33	Wilts	$\frac{1}{137}$
15	Gloucester	$\frac{1}{220}$	34	East Riding, York	$\frac{1}{130}$
16	Bucks	$\frac{1}{219}$	35	North Riding, York	$\frac{1}{127}$
17	Northampton } Warwick	$\frac{1}{217}$	36	Cumberland	$\frac{1}{118}$
			37	Salop	$\frac{1}{109}$
18	Nottingham	$\frac{1}{216}$	38	Hereford	$\frac{1}{108}$
			39	Westmoreland	$\frac{1}{89}$

WALES.

Number.	Counties.	Ratio	Number.	Counties.	Ratio
1	Glamorgan	$\frac{1}{393}$	6	Cardigan	$\frac{1}{152}$
2	Carnarvon	$\frac{1}{234}$	7	Denbigh	$\frac{1}{137}$
3	Anglesea } Brecon	$\frac{1}{203}$	8	Carmarthen	$\frac{1}{120}$
			9	Pembroke	$\frac{1}{94}$
4	Flint	$\frac{1}{169}$	10	Montgomery	$\frac{1}{87}$
5	Merioneth	$\frac{1}{103}$	11	Radnor	$\frac{1}{39}$

Appendix to the Second Annual Report of the Poor Law Commissioners.

" Dr. Casper has examined some peculiar circumstances which exercise an influence over the number of still-born children:—such as illegitimate conceptions, the abuse of strong drinks, &c. Thus at Goettingen, out of one hundred births, there were three legitimate still-born, and fifteen illegitimate. At Berlin, the still-born out of one hundred illegitimate births have been, during half a century, three times more numerous than out of the same number of legitimate; nor is this state of things at all ameliorated, for during the four years from 1819 to 1822, we have:—

	Living.	Dead.	One Death out of
Legitimate births ...	22.643	937	25 infants.
Illegitimate do.	4,002	317	12 do.

The Prussian official Tables give for 1827, (Bulletin de M. Ferrussac, January 1830, page 118,) quoted by M. Quetelet, 490,660 births, out of which 16,726 were still-born; or 1 out of 29.

" At Hamburgh, during the year 1829, in a house almost altogether devoted to diseased prostitutes, out of eighteen illegitimate births, six were born dead; and in another house in the same town, the receptacle also of prostitutes, out of ninety-three births, eleven were born dead.

" These different examples shew strongly the great influence which the condition of the mother has on the existence of her offspring, exhibiting the value of researches on this subject, on the causes which tend to multiply the number of still-born." *

---

* Quetelet, Vol. I. p. 136.

" The consideration of the mortality of infants naturally suggests an inquiry into the situation of the mothers. According to Willan, the mortality in the great Lying-in Hospitals, in London, where there are received annually nearly 5,000 females, has been—

	Mothers.	Infants.
From 1749 to 1758............	1 in 42	1 in 15
1759 .. 1768............	1 .. 50	1 .. 20
1769 .. 1778............	1 .. 55	1 .. 42
1779 .. 1788............	1 .. 60	1 .. 44
1789 .. 1798............	1 ..288	1 .. 77

" According to Dr. Casper, the mortality amongst the lying-in females in Berlin, has been—

From 1758 to 1763.....................	1 in 95
1764 .. 1774.....................	1 .. 82
1785 .. 1794.....................	1 .. 141
1819 .. 1822.....................	1 .. 152

" We see here how the mortality varies according to the care taken of the mother and infant during the accouchement, also according to the amount of scientific and practical knowledge in the medical department.

" The greatest mortality appears to have been in the Hotel Dieu, in Paris, at the end of the last century; it was 1 in 15 of the mothers, whilst in London it was reduced to 1 in 288, the mortality was therefore nineteen times less." *

---

* Quetelet, Vol. I. p. 130.

Note VII. Page 81.

Mr. T. R. Edmonds, in his " New Theory of the Causes producing Health and Longevity," observes that " there subsists the most intimate connexion between sickness and death; and, in the order of nature, the latter is preceded by the former as its cause. From a great extent of observations, I have collected the important fact, that death is proportional to *duration* of sickness alone, and is independent of intensity. These observations have been made on military masses of the greatest magnitude, under the widest variety of circumstances. They serve to establish the fact, that in any considerable number of men, placed for a given time under peculiar circumstances, there exists a fixed proportion between the number of deaths and the aggregate duration of sickness; and, what may appear extraordinary, the definite proportion which is applicable to one set of circumstances, agrees nearly with the definite proportion which is applicable to any other combination of circumstances. *Two years of sickness to each death* appears to be the law of nature, from which little deviation is allowed, except in very unhealthy climates. In the English army, at home and inactive, there are two years and a half of alleged sickness to each death. In the English West India army, there is one year and a third of sickness to each death.

" In the East Indies, the proportion more correctly stated, is two years and a third for the native troops, and one year and two-thirds for the European troops. The experience of benefit societies shews, that this proportion for the English working population approaches very near to *two years*. In any population between the ages of 20 and 55, if the numbers constantly sick amount to *four* per cent. on the living, then it may be safely inferred, that the annual deaths amount to *two* per cent. on the living."*

In the construction of tables for provision in sickness and in old age, Mr. Edmonds states that he has been influenced by the general principle, "that all savings from the earnings of labour ought to be made before the age of *fifty-five* years; that between the ages of 55 and 65 a man should expend the labour barely sufficient for his maintenance; and that for the portion of life which may be enjoyed after the age of 65, he should subsist entirely on previous savings."

NOTE VIII. Pages 88 and 90.

Savings' Banks have been very generally extended, and with the advantages that had been anticipated. In a poor country, and more especially in cities and towns however, perhaps the most valuable establishments are the " Small Loan Fund Societies," as supplying the strongest incentives to virtuous exertion. In the city of Cork, such an establishment exhibits, in the most striking manner, its beneficial influence. To Mr. J. M'Donnell, whose benevolence, activity, intelligence, and zeal in carrying out the principles of the society, and whose local knowledge in directing the early application of its funds, have entitled him to the thanks of his fellow citizens, I am indebted for the following information :—

This society commenced its operations in June, 1837, by the issue of debentures of 5*l.*, with a capital of only 65*l.*, while the loans amounted to 72*l.* the first week, the difference having been advanced by a benevolent individual. In six months the loans amounted to 279*l.*; at the end of the year to 550*l.*; and at the present time, after one year and a half's operation, to 600*l.* per week. 700 to 1,400 persons

* Edmonds. Chap. VIII.

pay instalments daily, being borrowers of sums from 5 shillings to 10 pounds.

Each applicant for a loan may be required to produce testimonials of character, and "must tender the joint and several promissory note of himself and two solvent securities, for the amount of the said loan."

Interest at the rate of *sixpence* in the pound, is deducted at the time of granting every loan, and every loan must be repaid by instalments of one shilling in the pound; the first payment to commence on a named day in the second week after borrowing. Should any instalments fail to be paid on the day it becomes due, the defaulters are fined one penny on every shilling, and should the defaulters extend to the next ensuing week, a further fine of two-pence on every shilling is inflicted, and immediate proceedings taken against the borrower and his securities, for what may be due on the loan.

It has been found that the capital subscribed, is capable of being turned seven times and a half in a year.

Some peculiarly interesting facts have been elicited in the working of this society; amongst many it may be stated, that no defaulter has been found where the sum borrowed was for the purpose of advancing his trade; and that the nearer the grade of the borrower and his securities, the more correctly has the interest and loan been paid.

Note IX. Page 90.

This subject is placed in a strong light by Mr. De Morgan. "No one," he says, "can form an accurate idea of such an establishment, (an Insurance Office,) who does not consider it as a Savings' Bank, yielding interest, and interest upon interest. This is the reason why an office, which charges for its insurance more than it is worth as an insurance, may, nevertheless, put its contributors in a better

position than they could have held if there had been no such
institution. To make this clear, let us consider the working
of a single investment office. A large number of individuals
subscribe a sum, which they intrust to an individual, or a
company to employ, yielding them the return at some fixed
but distant period. Let each share be 100*l.* The best
thing which an individual could do with such a small sum,
so as to have perfect security for its return, would be to invest it in the funds, at three and a half per cent. He might
also invest the interest, and thus obtain compound interest:
but it is not easy for an individual to do this; unless he provide an agent to draw the dividends immediately on their
becoming due, various circumstances will happen to prevent
the immediate investment of the interest. It is not at all an
unfair calculation to suppose that upon each half yearly
dividend a month will be lost, so that nominal compound
interest for 42 years, will only be really for 35 years. A
single pound, therefore, laid up by a man of 20, and im-
proved for the average term of his life, at three and a half
per cent. would only become 3*l.* one-third; while, in the
hands of a person who lost no time, it would become 4*l.* one-
fourth, or nearly a pound more. On the other hand, a
company, or a skilful individual who can command large
sums of money, can always make the best interest which
the market will afford. The funds, from the security of
their tenure, and the conveniences which they offer, will
always, in ordinary times, represent the lowest rate of interest which money will yield; other investments, which
offer better interest, are generally only accessible to those who
can command considerable sums, and are frequently attended
with risk: so that it requires knowledge to distinguish between the sound and the unsound. A company employing the whole time of a person or persons skilled in money
matters, and having continual large investments to make,
can realize not only more interest, but so much more, that

there shall remain a surplus worth considering, after the skill employed has been paid for."*

Note X. Page 95.

In Sir J. F. W. Herschel's "Preliminary Discourse on the Study of Natural Philosophy," a work which should be in the hands of all who set value on freedom of thought, elevation of sentiment, and the advancement of true knowledge, the following observations occur in connexion with this subject :—

" Not to trust the evidence of our senses, seems, indeed, a hard condition, and one which, if proposed, none would comply with. But it is not the direct evidence of our senses that we are in any way called upon to reject, but only the erroneous judgments we unconsciously form from them, and this only when they can be shewn to be so by counter evidence of the same sort; when one sense is brought to testify against another, for instance; or the same sense against itself, and the obvious conclusions in the two cases disagree, so as to compel us to acknowledge that one or other must be wrong. For example, nothing at first can seem a more rational, obvious, and incontrovertible conclusion, than that the colour of a body is an inherent quality, like its weight, hardness, &c., and that to see the object, and see it of its own colour, when nothing intervenes between our eyes and it, are one and the same thing.

" Yet this is only a prejudice ; and that it is so, is shown by bringing forward the same sense of vision which led to its adoption, as evidence on the other side; for when the differently coloured prismatic rays are thrown, in a dark room,

---

* De Morgan's " Essay on Probabilities," p. 239.

o

in succession upon any object, whatever be the colour we
are in the habit of calling its own, it will appear of the particular hue of the light which falls upon it: a yellow paper,
for instance, will appear scarlet when illuminated by red
rays, yellow when by yellow, green by green, and blue by
blue rays; its own (so called) proper colour not in the least
degree mixing with that it so exhibits.

" To give one or two more examples of the kind of illu-
sion which the senses practise on us, or rather which we
practise on ourselves, by a misinterpretation of their evidence : the moon, at its rising and setting, appears much
larger than when high up in the sky. This is, however, a
mere erroneous judgment; for when we come to measure its
diameter, so far from finding our conclusion borne out by
fact, we actually find it to measure less. Here is eyesight
opposed to eyesight, with the advantage of deliberate
measurement. In ventriloquism, we have the hearing at
variance with all the other senses, and especially with the
sight, which is sometimes contradicted by it in a very extraordinary and surprising manner, as when the voice is
made to seem to issue from an inanimate and motionless
object. If we plunge our hands, one into ice-cold water,
and the other into water as hot as can be borne, and after
letting them stay a while, suddenly transfer them both to a
vessel full of water at a blood heat, the one will feel a
sensation of heat, the other of cold. And if we cross the
two first fingers of one hand, and place a pea in the fork
between them, moving and rolling it about on a table, we
shall (especially if we close our eyes) be fully persuaded
we have two peas. If the nose be held while we are eating
cinnamon, we shall perceive no difference between its flavour
and that of a deal shaving.

" These, and innumerable instances we might cite, will
convince us, that though we are never deceived in the

sensible impression made by external objects on us, yet in
forming our judgments of them, we are greatly at the mercy
of circumstances, which either modify the impressions actually received, or combine them with adjuncts which
have become habitually associated with different judgments;
and, therefore, that in estimating the degree of confidence
we are to place in our conclusions, we must, of necessity,
take into account these modifying or accompanying circumstances, whatever they may be."*

Note XI. Page 95.

" When two circumstances happen to change together,"
observes Mr. De Morgan, " it is frequently presumed that
they are connected with each other, when, in truth, there
is no reason for any such supposition.   An individual who
has been unlucky during several games, happens to begin
to win after the introduction of new cards.   His fortune
changes, as most probably it will do ; for, if the chances be
even, and three games have previously been lost, it is seven
to one against the next three games resembling them, and
an even chance that he shall win the next game.   If he win
the next, or if, indeed, he does not go on losing, he notes
the circumstance, and the next time a run of ill luck occurs,
he takes particular care to repeat the experiment.   In this
way he soon furnishes himself with a tolerable number of
facts in support of his theory.   The exceptions are for-
gotten ; for it is the character of negative events, to lay less
firmly hold of the mind than positive ones.   The lucky hit
of a prophet of the weather, in foretelling the coldest day of
January, 1838, did more to establish his infallibility, than

* Preliminary Discourse, p. 81.

weeks of succeeding mistakes could destroy. Thus the theory of the change of the weather with that of the moon, receives more confirmation from one fact, than of doubt from two against it. It was frequently supposed, a few years ago, that comets produced hot weather. An examination of the number of comets discovered in years of different average temperature, gave it as a result that there were more comets in hot summers than in cold ones. But since hot summers are generally fine, with clear skies, and cold summers cloudy and rainy, it is obvious that the former are more favourable to the discovery of comets than the latter. The fact, then, from which the inference was drawn, amounted to this, that the years of heat are those in which *we see* most comets. With what we know of the matter, there is no more reason to suppose that comets bring heat than that heat brings comets. We must, in all instances of presumed connexion, look closely at these two distinct things, the happening of an event, and our perception of it; otherwise we shall be always liable to suppose that an event may produce the first, when it produces only the second." *

Note XII. Page 103.

Mr. Quetelet, in his work " on Man," &c., enters into farther detail. " During the four years," he says, " that preceded 1830, 28,686 persons were accused before the Courts of Assize, in France, or annually about 7,171 individuals: this gives one accused person for 4,463 inhabitants, taking the population at 30,000,000 of souls. Farther, out of 100 accused, 61 were condemned to punishments more or less severe.

---

* De Morgan's " Essay on Probabilities," p. 120.

The following Table presents the result of each.

Years	Accused.	Con-demned.	Inhabitants for one accused.	Con-demned of 100 accused.	Crimes		Ratio between the two sorts of crime.
					Against the person.	Against the property	
1826	6,988	4,348	4,557	62	1,907	5,081	2·7
1827	6,929	4,236	4,593	61	1,911	5,018	2·6
1828	7,396	4,551	4,307	61	1,844	5,552	3·0
1829	7,373	4,475	4,321	61	1,791	5,582	3·1
Total	28,686	17,610	4,444	61	7,453	21,233	2·8
1830	6,962	4,130	4,576	59	1,666	5,296	3·2
1831	7,607	4,098	4,281	54	2,046	5,560	2·7
Mean	7,284	4,114	4,392	56	1,856	5,428	2 9

The activity of the judicial police was necessarily relaxed during the latter months of 1830, in consequence of the political condition of the country, so that judgments in many cases belonging to that period, were deferred to 1831, and thereby increased the figures for that year.

The following Table relates to the state of crime in Belgium for a similar period:—

Years	Accused.	Con-demned.	Inhabitants for one accused.	Con-demned for 100 accused.	Crimes		Ratio.
					Against the person.	Against the property	
1826	725	611	5,211	84	189	536	2·8
1827	800	682	4,776	85	220	580	2·6
1828	814	677	4,741	83	230	584	2·5
1829	753	612	5,187	81	203	550	2·7
1830	741	541	5,274	73	160	581	3·8
Mean	767	625	5,038	82	200	566	2·8

Quetelet "Sur l'Homme," Vol. II. p. 185.

From the returns made to the British Parliament of the number of criminal offenders in England and Wales, and Ireland in 1837, we find the following results:—

	Accused.	Con-victed.	Not guilty on trial.	No bills found.	No prosecu-tion.	Total.	Con-demned of 100 accused.
England and Wales. .	23,612	17,096	4,388	1,637	471	6,496	72·4
Ireland ..	14,804	9,536	3,011	1,333	906	5,268	64·4

" Amongst all the causes," says M. Quetelet, " which influence the development or retardation of the disposition to crime, the most energetic is, without doubt, the age. It is, in effect, with the age that the physical force and the passions of man are developed, and subsequently decrease ; it is also with the age that his reason is developed, that it continues to increase till his power and passions have passed their maximum of intensity. By considering only the three elements, the physical force, the passions, and the reason of man, it may be almost declared *à priori,* what ought to be the degrees of disposition to crime at different ages. This disposition ought to be nearly null at the two extremes of life ; on the one side physical force and the passions, those two powerful instruments to crime, have scarcely entered into life, and on the other side, their energy, nearly extinguished, is altogether destroyed by the influence of reason : the disposition to crime, on the contrary, ought to be at its *maximum* at the age when the physical forces and the passions have attained their *maximum,* and where reason had not acquired sufficient empire to control their combined influence.

The following Table shews the number of crimes against the person, and against property, which had been committed in France, by the two sexes, during the years 1826, 27, 28, and 29, with the ratio of these numbers; the fourth column shews how a population of 10,000 souls is divided, in France, according to the ages; and the last column shews the ratio of the total number of crimes to the number corresponding to the preceding column.

Ages.	Crimes		Crimes against property out of 100 crimes.	Population according to age.	Degree of disposition to crime.
	Against the person.	Against the property			
Less than 16	80	440	55	3,304	161
16 to 21	904	3,723	80	887	5,217
21 .. 25	1,278	3,329	72	673	6,846
25 .. 30	1,575	3,702	70	791	6,671
30 .. 35	1,153	2,883	71	732	5,514
35 .. 40	650	2,076	76	672	4,057
40 .. 45	575	1,724	75	612	3,757
45 .. 50	445	1,275	74	549	3,133
50 .. 55	288	811	74	482	2,280
55 .. 60	168	500	75	410	1,629
60 .. 65	157	385	71	330	1,642
65 .. 70	91	184	70	247	1,113
70 .. 80	64	137	68	255	788
80 and above	5	14	74	55	345

Quetelet, " Sur l'Homme," Vol. II. p. 240.

" It will be seen that man begins to exercise his disposition for crime first against property; from 25 to 30 years, when his powers are developed, it is against the person, and that towards the age of 25 years the disposition to crime attains its *maximum.*"

The following Table from the Parliamentary Return for 1837, shews the number in that year at the different periods of life, and a comparison of the proportion at these ages, in each of the three last years.

	Males.	Females	Total.	Ratio per 100 1837.	Ratio per 100 1836.	Ratio per 100 1835.
Aged 12 and under	303	55	358	1·52	1·84	1·67
12 to 16....	1,962	334	2,296	9·72	9·71	9·70
16 .. 21....	5,774	1,128	6,902	29·23	29·03	29·65
21 .. 30....	6,172	1,322	7,494	31·74	31·42	31·92
30 .. 40....	2,758	681	3,439	14·56	14·43	1·401
40 .. 50....	1,211	360	1,571	6·65	6·76	6·60
50 .. 60....	587	177	764	3·24	3·33	3·24
Above 60....	284	81	365	1·55	1·40	1·30
Not ascertained ..	356	67	423	1·79	2·08	1·91
	19,407	4,205	23,612			

To this I add the following Table, formed from the Six-teenth Report of the Inspectors General of Prisons in Ireland.

	Males.	Females.	Total.	Ratio per 100.
Aged 12 and under ..	86	21	107	0·7
12 to 16......	765	150	915	6·2
16 .. 21......	2,460	835	3,295	22·3
21 .. 30......	4,413	1,424	5,837	39·4
30 .. 40......	1,975	604	2,579	17·4
40 .. 50......	889	275	1,164	7·9
50 .. 60......	273	95	368	2·5
Above 60......	104	39	143	0·9
Age not ascertained..	355	41	396	2·7
	11,320	3,484	14,804	

The greater number of these offences are thus classed :—

ENGLAND AND WALES.				
	Males.	Females.	Ratio of one crime to the other.	
			Males.	Females
Of those against the person ..	1,538	181	—	—
Against property, with violence	1,329 ⎱ 16,560	71 ⎱ 3,724		
Without violence	15,231 ⎰	3,653 ⎰	10·77	20·57

IRELAND.				
Of those against the person ..	2,698	434	—	—
Against property, with violence	606 ⎱ 5,023	56 ⎱ 2,602		
Without violence	4,417 ⎰	2,546 ⎰	1·86	6·00

The population being taken at 14,000,000 for England and Wales, there are found to be 593 inhabitants for one accused person.

The population being taken at 8,000,000 for Ireland, there are found to be 540·5 inhabitants for one accused person.

If, however, we distinguish the nature of the crimes, we shall find a wide difference between the conditions of the two countries.

	Inhabitants for one accused.		Ratio of one crime to the other.
	Against the person.	Against property.	
England and Wales......	8,144	548	14·86
Ireland	2,554	1,049	2·43

Hence, crime against property is, in England, nearly 15 times greater than that against the person, while in Ireland it is only 2½ times greater.

The statistical documents to which we have referred, afford also some interesting facts relative to the influence which education exercises on the tendency to crime.

Of the 23,612 persons accused in England and Wales, those who possessed—

	Males.	Females	Ratio per 100.	
			Males.	Females.
Superior instruction	98	3	0·50	0·07
Could read and write well	2,057	177	10·61	4·22
Read and write imperfectly....	10,147	2,151	52·28 ⎫ 86·72	51·15 ⎫ 93·48
Neither read nor write	6,684	1,730	34·44 ⎭	42·33 ⎭
Instruction not ascertained	421	94	2·17	2·23
	19,407	4,205		

Taking together those who could neither read nor write, and who could read and write imperfectly, we have, of males 86·72 per 100, and of females 93·48 per 100.

Of the 14,804 persons accused in Ireland, those who could—

	Males.	Females	Ratio per 100.	
			Males.	Females.
Read and write....	4,020	491	35·51	14·00
Read only	2,170	886	19·17 ⎱ 57·92	25·43 ⎱ 81·40
Neither read nor ⎱ write .... ⎰	4,386	1,950	38·75	55·97
Instruction not as- ⎱ certained. .... ⎰	744	157	6·57	4·50
	11,320	3,484		

M. Quetelet supplies us with a similar Table for France.

Intellectual state of the accused.	1828 and 1829.	Ratio per 100.	1830 and 1831.	Ratio per 100.
Neither read nor ⎱ write .... ⎰	8,689	60·80 ⎱ 87·43	8,919	61·23 ⎱ 87·82
Read and write ⎱ imperfectly . .. ⎰	3,805	26·63	3,873	26·59
Read and write ⎱ well. .... ⎰	1,509	10·57	1,455	9·99
Superior instruc- ⎱ tion. .... ⎰	286	2·00 ......·	319	2·19
	14,289		14,566	

The number of accused in the superior class, was augmented in 1830 and 1831, in consequence of the political state of that period.

We may observe, that in Ireland the acquisition of a

knowledge of letters can not, as in England, be supposed to mark an elevation of moral condition. Education there appears to have been commenced at the wrong end. Habits of order, of respect for the law, and for those whom the law has placed in authority, should, according to the declaration of nature herself, in the gradual development of man, have precedence, if permanent improvement is to be expected in a people. The remarkable fact recorded by Mr. Brian, in his "Practical View," p. 194, affords a melancholy example of this strange inversion :—" Out of 1,400 adult mendicants in Dublin," he says, " who were examined for that purpose, 600 were found capable of reading, and that distinctly." A proportion which is not equalled, we believe, by the population of any European country.

Note XIII. Page 107.

" We must never forget," says Sir J. F. W. Herschel, " that it is the principles, not phenomena,—laws, not insulated, independent facts, which are the objects of inquiry to the natural philosopher. As truth is single, and consistent with itself, a principle may be as completely and as plainly elucidated by the most familiar and simple fact, as by the most imposing and uncommon phenomenon. The colours which glitter on a soap-bubble, are the immediate consequences of a principle the most important, from the variety of phenomena it explains, and the most beautiful, from its simplicity, and compendious neatness, in the whole science of optics. If the nature of periodical colours can be made intelligible by the contemplation of such a trivial object, from that moment it becomes a noble instrument in the eye of correct judgment: and to blow a large, regular, and durable soap-bubble, may become the serious and praise-worthy endeavour of a sage, while children stand round and scoff, or children of a larger growth hold up their hands in astonishment, at such waste of time and trouble.

" To the natural philosopher there is no natural object un-important or trifling. From the least of nature's works he may learn the greatest lessons. The fall of an apple to the ground, may raise his thoughts to the laws which govern the revolutions of the planets in their orbits; or the situa-tion of a pebble may afford him evidence of the state of the globe he inhabits, myriads of ages ago, before his species became its denizens."*

Note XIV. Page 108.

Daughter of Cælus and Terra. Her oracle, in Attica, was considered very famous in the time of Deucalion, who was said to have consulted it relative to the loss of man-kind. She is represented with a sword in one hand and a pair of scales in the other— Ovid's Metamorphosis.

* Herschel's Preliminary Discourse, p. 14.

P